Brazilian Marine Biodiversity

Series Editor

Alexander Turra, Ocean Sustainability, Oceanographic Institute and Advanced Studies Institute
University of São Paulo
São Paulo, Brazil

The book series Brazilian Marine Biodiversity was designed to communicate to a broad and international readership the diversified marine and coastal habitats along the large Brazilian coast.

The diversity of marine habitats found in Brazil is astonishing and includes estuaries, coral reefs, rocky shores, sandy beaches, rhodolith beds, mangroves, salt marshes, deep-sea habitats, vegetated bottoms, and continental shelf. These habitats are addressed from an ecosystem perspective across the series, and characterized in terms of distribution and peculiarities along the Brazilian coast, records of relevant species, and information on the prevailing structuring ecological and oceanographic processes governing biodiversity.

This series is an initiative of the UNESCO Chair for Ocean Sustainability, hosted by the Oceanographic Institute and the Advanced Studies Institute of the University of São Paulo, within the Brazilian Network for Monitoring Coastal Benthic Habitats (ReBentos; rebentos.org), which is supported by the Brazilian National Council for Scientific and Technological Development (CNPq), the Research Program on Biodiversity Characterization, Conservation, Restoration and Sustainable Use of the São Paulo Research Foundation (BIOTA-FAPESP), the Coordination for the Improvement of Higher Education Personnel (CAPES) and the Brazilian Innovation Agency (FINEP). ReBentos is part of the Brazilian Network on Global Climate Change Research (Rede Clima) and the Science and Technology National Institute on Climate Changes (INCT Mudanças Climáticas) at the Ministry of Science, Technology, and Innovation (MCTI).

Ricardo Coutinho • Ronaldo A. Christofoletti
Editors

Brazilian Rocky Shores

Springer

Editors
Ricardo Coutinho
Instituto de Estudos do Mar Almirante
Paulo Moreira
Departamento de Biotecnologia Marinha
Arraial do Cabo, RJ, Brazil

Ronaldo A. Christofoletti
Instituto do Mar
Universidade Federal de Sao Paulo
(IMar/UNIFESP)
Santos, SP, Brazil

ISSN 2520-1077 ISSN 2520-1085 (electronic)
Brazilian Marine Biodiversity
ISBN 978-3-031-67205-7 ISBN 978-3-031-67206-4 (eBook)
https://doi.org/10.1007/978-3-031-67206-4

This Springer imprint is published by the registered company Springer Nature Switzerland AG
The registered company address is: Gewerbestrasse 11, 6330 Cham, Switzerland

About the Book

Rocky shores correspond to a very particular habitat in Brazil, which is concentrated in the Southeastern and Southern coasts. There are different types of substrata and seascapes that cause a high environmental heterogeneity responsible to host a high species biodiversity. Primary and secondary production, as well as structuring organisms and processes, differs from temperate areas.

The objective of the rocky shores volume is to bring an overview of current knowledge about this habitat to the entire international community. More than a scientific review of published articles, this is also a space for the accumulated knowledge of researchers working in this environment to be compiled in this book.

The book aims to provide accessible language not only to researchers but also to undergraduate students and other sectors of society that can benefit from this publication. The book focuses on consolidated, intertidal, and shallow sublittoral substrates. It should be noted that there will be a special volume on coral reefs, so this topic may be mentioned, but should not be treated in depth in the chapters of this volume.

Contents

Contributors

Luciana Altvater Instituto de Estudos do Mar Almirante Paulo Moreira, Departamento de Biotecnologia Marinha, Arraial do Cabo, RJ, Brazil

Juliana Imenis Barradas Center for Natural and Human Sciences, Federal University of ABC, Postgraduate Program in Teaching and History of Science and Mathematics (PEHFCM), São Paulo, Brazil

Municipal University of São Caetano do Sul, São Paulo, Brazil

Leandro Calado Instituto de Estudos do Mar Almirante Paulo Moreira, Departamento de Biotecnologia Marinha, Arraial do Cabo, RJ, Brazil

Ronaldo A. Christofoletti Universidade Federal de São Paulo (UNIFESP), Instituto do Mar, Santos, SP, Brazil

Mariana Clauzet National Institute of Science and Technology in Public Policies, Strategies and Development (INCT-PPED), Rio de Janeiro, Brazil

Cesar A. M. M. Cordeiro Universidade Estadual do Norte Fluminense Darcy Ribeiro, Laboratório de Ciências Ambientais, Campos dos Goytacazes, RJ, Brazil

Ricardo Coutinho Instituto de Estudos do Mar Almirante Paulo Moreira, Departamento de Biotecnologia Marinha, Arraial do Cabo, RJ, Brazil

Bernardo A. P. da Gama Universidade Federal Fluminense, Departamento de Biologia Marinha, Niterói, RJ, Brazil

Frederico T. S. Tâmega Instituto de Estudos do Mar Almirante Paulo Moreira, Departamento de Biotecnologia Marinha, Arraial do Cabo, RJ, Brazil

Lohengrin D. A. Fernandes Instituto de Estudos do Mar Almirante Paulo Moreira, Departamento de Biotecnologia Marinha, Arraial do Cabo, RJ, Brazil

Carlos E. L. Ferreira Laboratório de Ecologia e Conservação de Ambientes Recifais, Universidade Federal Fluminense, Niterói, RJ, Brazil

Sergio R. Floeter Universidade Federal de Santa Catarina, Laboratório de Biogeografia e Macroecologia Marinha, Florianópolis, SC, Brazil

Natalia Pirani Ghilardi-Lopes Center for Natural and Human Sciences, Federal University of ABC, Postgraduate Program in Teaching and History of Science and Mathematics (PEHFCM), São Paulo, Brazil

Stuart R. Jenkins School of Ocean Sciences, Bangor University, Menai Bridge, Anglesey, UK

Aline Sbizera Martinez Instituto do Mar, Universidade Federal de São Paulo, Santos, SP, Brazil

Sérgio A. Navarrete Estación Costera de Investigaciones Marinas and Center for Advanced Studies in Ecology and Biodiversity (CASEB), Departamento de Ecología, Facultad de Ciencias Biológicas, Pontificia Universidad Católica de Chile, Santiago, Chile

André Pardal Universidade Federal de São Paulo (UNIFESP), Instituto do Mar, Santos, SP, Brazil

Valéria Vinha Graduate Program in Public Policies, Strategies and Development at Federal University of Rio de Janeiro, Institute of Economics (PPED/IE/UFRJ), Rio de Janeiro, Brazil

Nelson Violante-Carvalho Universidade Federal do Rio de Janeiro, Programa de Engenharia Oceânica, Reitoria—Cidade Universitária, Rio de Janeiro, RJ, Brazil

Kringpaka Wangkulangkul Division of Biological Science, Faculty of Science, Prince of Songkla University, Songkhla, Thailand

Ilana Rosental Zalmon Universidade Estadual do Norte Fluminense Darcy Ribeiro, Laboratório de Ciências Ambientais, Campos dos Goytacazes, RJ, Brazil

About the Editors

Ricardo Coutinho graduated in Biological Sciences in the University of Iguaçu, UNIG (1979) and completed MS in Biological Oceanograph at the Federal University of Rio Grande (FURG) (1982), PhD in Biology (Botany and Ecology) at the University of South Carolina (1987), and Post-doc at Duke Marine Lab (2003) and Woods Hole Oceanograph Institute (2004), USA. Since 1987 he is a researcher at the Instituto de Estudos do Mar Almirante Paulo Moreira (IEAPM). He is interested in marine ecology and in studying the impacts of invasion species and local stressors on marine biodiversity. **E-mail:** rcoutinhosa@yahoo.com

Ronaldo A. Christofoletti graduated in Biological Science in the University of São Paulo (1999) and completed MS in Comparative Biology (2002) and PhD in Zootecnia, Universidade Estadual Júlio de Mesquita Filho (UNESP) (2025), and Post-doc at University of Wales, Bangor, UK (2010). Since 2014, he is a Professor at Universidade Federal de São Paulo (UNESP). Since 2012, he is Professor at the Institute of Marine Science in Universidade Federal de São Paulo (UNESP). He is interested in Marine Ecology, Scientific Communication and Ocean Literacy. **E-mail:** ronaldochristofoletti@gmail.com

Chapter 1
Brazilian Rocky Shores: Geomorphology and Physical Environment

Leandro Calado, Nelson Violante-Carvalho, and Frederico T. S. Tâmega

Abstract Rocky shores are formations of crystalline rocks that are close to the sea and are susceptible to weather, variations in the average sea level, sea currents, rains, and winds and are represented under different formations such as boulders, cliffs, and rocky shores with different degrees of slopes and extensions and may have terraces of abrasion by waves and fallen rocks. Erosion by wave action, wind, and rain are the main factors; however, temperature also influences the formation and composition of rocks.

Rocky Shores

The benthic marine coastal environments are among the most productive on the planet (Coutinho 2004). Among the ecosystems present in the intertidal region and habitats in the coastal zone, rocky shores are one of the most important because many species with ecological and economic importance live around it (Andrew and Mapstone 1987). Due to their location in the transition between the terrestrial and marine environment, rocky shores receive a large input of nutrients from terrestrial systems, which increases biomass and primary production of microphytobenthos and marine macroalgae (Coutinho 2004). This makes rocky shores a local habitat for feeding, growth, and reproduction of a large number of species, and due to their location it makes rocky shores one of the most popular and well-studied marine ecosystems. However, they are also threatened by inappropriate human use and occupation (Coutinho 2004).

Rocky shores are formations of crystalline rocks that are close to the sea and are susceptible to weather, variations in the average sea level, sea currents, rains, and

L. Calado (✉) · F. T. S. Tâmega
Instituto de Estudos do Mar Almirante Paulo Moreira, Departamento de Biotecnologia Marinha, Arraial do Cabo, RJ, Brazil

N. Violante-Carvalho
Programa de Engenharia Oceânica, Universidade Federal do Rio de Janeiro, Av. Pedro Calmon, Rio de Janeiro, Brazil

R. Coutinho, R. A. Christofoletti (eds.), *Brazilian Rocky Shores*, Brazilian Marine Biodiversity, https://doi.org/10.1007/978-3-031-67206-4_1

winds and are represented under different formations such as boulders, cliffs, and rocky shores with different degrees of slopes and extensions and may have terraces of abrasion by waves and fallen rocks (Suguio 2001; Coutinho 2004). The rocky shores suffer the actions of physical, chemical, and biological agents (Suguio 2001). In physical agents, erosion by wave action, wind, and rain are the main factors; however, temperature also influences the formation and composition of rocks through the expansion and contraction of minerals (Suguio 2001). The chemical agents depend on the composition of the rock since the minerals react chemically with sea water in addition to climatic factors (Suguio 2001). Regarding the actions of biological agents, we highlight the bioerosion promoted by organisms such as sea urchins, barnacles, and mollusks (Pappalardo et al. 2018).

The rocky shores are important for the evolutionary history of the coastal zone, as they are testimonies of the variations in the average sea level that occurred during in the past (Suguio 2001). The current morphology of the rocky shores can provide erosive indicators of the variations in the average sea level, recorded as platforms of marine abrasion arranged in several levels, above or below the current average sea level (Vilano and Souza 2011). Multidisciplinary studies (biology, paleontology, oceanography, geology, geomorphology, and hydrodynamics) of rocky shores can provide important past and current records for assessing these environments, as well as assist in decision-making for the impacts of climate change and its effects, in addition to subsidize legal instruments for their conservation and protection.

Along the Brazilian coast, the real rocky shores are present, almost exclusively, in the Southeast and South regions, where the coastal plateau reaches the coastline, forming an irregular coast supported by rocky promontories and segmented into islands (Coutinho 2004; Villwock et al. 2005). The main features of the Southeastern region are the proximity of Serra do Mar which, in many points, reaches directly to the sea (Coutinho 2004). The sudden inflection to the West, in the Cabo Frio region, the orientation of the coast and the bathymetry, due to the fracture zone in Rio de Janeiro, caused the structural alignments of the crystalline basement, from Northeast to Southwest, to be truncated by the approximately East-West orientation of the coast, between Cabo Frio and Angra dos Reis Bay (Coutinho 2004).

Alto de Cabo Frio is an offshore and onshore structure with Northeast-Southeast orientation formed during the process of breaking and separating of Gondwana and opening up the South Atlantic more than 130 million years ago. It is characterized by a high incidence of magmatic rocks associated with sedimentary volcanic rocks. This important magmatism is related to central volcanic duct and slightly alkaline composition. The youngest rocks that occur in the region are related to the formation process of Alto de Cabo Frio and are represented by two main igneous units: meso-cenozoic diabase dykes (with ages varying between 130 Ma and 55 Ma) and dykes and Alkaline tertiary plugs (with ages ranging from 93 to 52 Ma).

The main part of the Brazilian coast, where rocky shores are among the most important ecosystems, comprises Cabo Frio (Rio de Janeiro State) to Cabo de Santa Marta (Santa Catarina State). In this region, with the exception of the area influenced by the upwelling of Cabo Frio, which has benthic fauna and flora with temperate to tropical affinities and acts as a biogeographic barrier for a large number of species, the specific composition is of a subtropical characteristic, with a high diversity of species (Coutinho 2004).

Based on the physical-environmental units suggested by the regional coordinations, we can identify the following units that have significant rocky shores on the Brazilian coast (Coutinho 2004). Not considered here are the sandstone reefs and rock formations common on the Northeast coast.

- Pinhal (RS) up to the border of Rio Grande do Sul/Santa Catarina States
- Border of Rio Grande do Sul/Santa Catarina States up to the border of Laguna/ Jaquaruna (Santa Catarina State)
- Border of Laguna/Jaquaruna to Ponta da Faísca or Gamboa (Santa Catarina State)
- Ponta da Faísca or Gamboa to Ponta dos Ganchos (border of Governador Celso Ramos/Tijuca (Santa Catarina State)
- Ponta dos Ganchos to Barra do Sul (Santa Catarina State)
- Barra do Sul (SC) limit of States Paraná/Santa Catarina
- Limit of Paraná State to Pontal do Sul (Paraná State)
- Pontal do Sul (PR) to Peruíbe, limit with Juréia (São Paulo State)
- Peruíbe, limit with Juréia to Boracéia beach (São Paulo State)
- East of Boracéia beach to West of Restinga de Marambaia (Rio de Janeiro State)
- West of Restinga de Marambaia to Cabo Frio (Rio de Janeiro State)
- Cabo Frio to Paraíba do Sul River (Rio de Janeiro State)
- Paraíba do Sul River to left margin of Piraquê River (Espírito Santo State)
- Right margin of the Piraquê River up to the limit of the State of Espírito Santo/Bahia

Hydrodynamic Processes

A Southern and Southeastern brazilian continental margin, where most of the rocky shore environments of Brazil are unique regions. The physical phenomena that occur in this region of the continental shelf (CS) are, for the most part, independent of the dynamics of the deep ocean. The CS is the most coastal part of the continental margin, averaging around 200 km in width, between the coast and the shelf break, generally occurring at a depth of 200 m along the world's continents. The presence of the coast is important to intensify the occurrence of divergences and convergences. The small depth, compared to the average of the oceans, accentuates the vertical shear. Its small volume makes it more prevalent as ocean-atmosphere exchanges, making seasonal processes more intensified. Its horizontal anisotropy, given by the marked difference between length and width, has very important implications for how dynamic processes occur.

According to Castro (1996), the CS can be divided into three regions: internal, medium, and external. The internal is where there is interaction with the coast and the external makes the communication between the dynamics of the geostrophic currents, such as the Brazil Current (BC) and the dynamics of the CS. In the

South-Southeast region, we can subdivide a CS between the Campos Region, extending from Arraial do Cabo and Cabo de São Tomé to the Vitória region. The coastline in this portion of the CS is oriented approximately from Southeast to Northeast, with a width that varies between 70 km and 100 km. The Cabo Frio region, where coastline direction changes direction to East-West, with depths up to the shelf break between 140 m and 200 m and width between 70 km and 150 km from the coast until the shelf break (Castro 1996) and extends to southern Brazil.

The processes that occur in the CS are essentially forced by the action of the wind in different time and space scales, tides, sea-surface elevation gradients and density fronts, friction, and geostrophic currents. According to Castro (1996): "*Tides represent the ocean's response to astronomical gravitational forces. On word continental shelves, however, sea-level oscillations and currents at astronomical tidal frequencies are generally seen as co-oscillations with the deep ocean. That is, most of the energy is with the waves excited in the coastal region by global oscillations in the deep ocean, with communication being established through the slope and the breaking of the continental shelf.*"

The wind in the CS has its most important role when it is acting parallel to the coast, with the acceleration of the Coriolis effect, which is a consequence of the Earth's rotation. The wind is capable of generating drift transports in the direction perpendicular to the coast. The result is a change in sea-level rise at the edges of the CS, generating geostrophic currents as a result. Thus, the internal and medium CS has its dynamics directly linked to the wind direction acting on the CS. These currents in the CS are responsible for the transport of organisms, which are often the species of the population of the rocky shore along the coast. On the other hand, the outermost part of the CS is almost entirely dependent on the dynamics of the deep ocean, which in the case of the southeastern and southern regions, consists of the presence of the BC and its mesoscale activity. However, in some regions, BC has direct influence and in others, its presence does not influence the dynamics of CS. The specific regions of Cabo Frio and Arraial do Cabo, for example, have unique current dynamics on the Brazilian coast. Despite being a coastal region, its rocky shores are influenced by the dynamics of typical external CS currents with a strong influence from the inner edge of the BC.

The Brazil Current is an intense jet that transports a considerable amount of water off the southeast of Brazil. This current is responsible for transporting tropical water (TA) at levels of 0–200 m and South Atlantic Central Water (SACW) at levels of 200–400 m (Silveira et al. 2000). Centered between the 200 and 1000 m isobaths, BC occupies the external part of the Southeastern Brazilian continental shelf and the continental slope (Castro 1996). BC has an intense mesoscale activity off the southeast, with large meanders that are observed between Vitória (20°S) and Cabo de Santa Marta (CSM, 28.5°S). Those eventually close in eddies and are advected, absorbed, or emitted by the BC (Mascarenhas et al. 1971; Silveira et al. 2000; Calado et al. 2008). In the South-Southwest region, there are important locations described in the literature as having recurring meandering characteristics: Vitória, Cabo de São Tomé (CST, 22°S), Cabo Frio (CF, 23°S), and Cabo de Santa Marta Grande (28.5°S). Cyclonic eddies are believed to be able to interact with the coastal part. Lee and Pietrafesa (1987) studied the relationship between Gulf Stream (GS) variability and coastal phenomena on the southeastern CS in North America. These authors found that three important oceanographic conditions occurred simultaneously so that, for example, the coastal upwelling become mature in that region, that is, with the thermocline rising on the surface. They are the front position of the GS in the region of the external shelf, the occurrence of cyclonic meanders, and the wind stress in the north direction.

Upwelling is understood as the rise of the thermocline on the sea surface. Coastal upwelling, therefore, is the emergence of the thermocline near the coast. The physical process responsible for this outcrop is the divergence of Ekman's transport on the coast, caused by the action of winds that induce transport perpendicular to the coast toward the open ocean, within the superficial ocean limit layer. In order for the water volume to be replaced, there is a response from the ocean characterized by a return transport toward the coast, below the superficial limit layer. This return transport, therefore, has the capacity to bring toward the coast colder waters that are in deeper regions, far from the surface (Calado et al. 2018). As an example of upwelling location of important rocky shores, Fig. 1.1 illustrates the temperature variation at two points with thermometers on rock shores over approximately 2.5 years (May 2017 to January 2020) at Arraial do Cabo region. The time series allows to identify the exposure to temperatures below 18 °C throughout the year, which is typical of the upwelling. The populations that inhabit rocky shores are directly impacted by the seasonality of temperatures.

In Fig. 1.2, it is possible to observe, with a satellite image, the influence of the cold-water plume formed in upwelling events at CF region and its influence up to Rio de Janeiro. This cold plume, rich in nutrients, occurs in all parts of the coast where conditions are favorable. When it reaches the photic zone, it "fertilizes" the region promoting an environment rich in organisms in those regions.

Coastal upwelling occurs in different parts of the world ocean. In general, on the west side of the continents, coastal upwelling occurs almost permanently and over large areas along the coast. These processes occur because the orientation of the coastline and the winds in these locations are favorable to the occurrence of the phenomenon, practically throughout the year. We can cite examples from the coast of Chile, Peru, California, and the West coast of Africa. In the East coasts of the continents, in general, the wind does not act with a component parallel to the coast strong enough to promote upwelling, except in some typical regions with regional characteristics that coincide for the occurrence of the phenomenon. Along the southeastern Brazilian CS, there are some locations that on certain occasions become favorable to coastal upwelling. In the Brazilian coast, upwelling is observed in several locations, typical of rocky shore environments: as in Vitória, Campos dos Goytacazes, Cabo Frio, Rio de Janeiro, Sepetiba, São Sebastião, Ubatuba, and Cabo de Santa Marta; however, Cabo Frio is notably the most intense. The phenomenon tends to occur at times of the year when the NE winds typically blow more frequently and with greater intensity (spring and summer) and tends to be less common in the autumn and winter months, when the passage of frontal systems is more frequent, where reversing the direction of the wind and promoting the opposite process, that is, coastal subsistence.

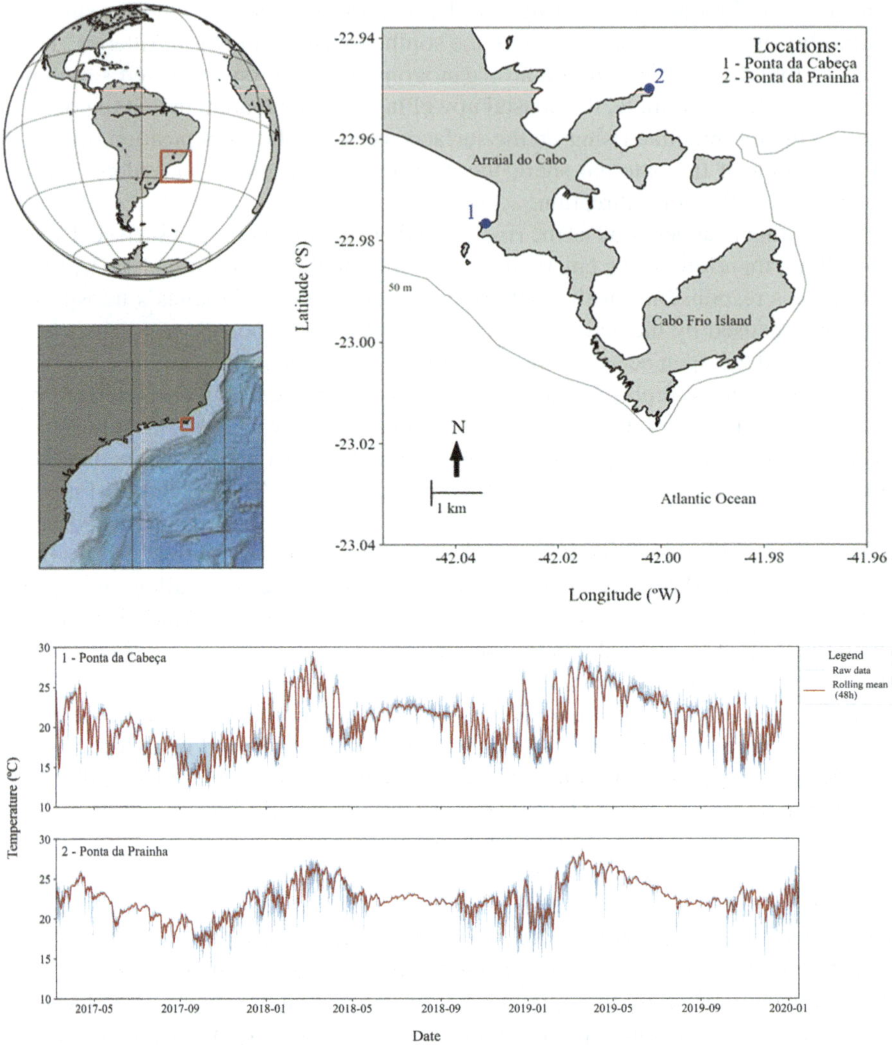

Fig. 1.1 Region of rocky shores of Arraial do Cabo. On the left, the location of the points where thermometers were installed on the rocks of Ponta da Cabeça (1) and Ponta da Painha (2). On the right, the temperature evolution between May 2017 and January 2020 at these two points. Note that the region has a clear influence of cold waters typical of the coastal upwelling of the region, especially in the spring and summer months (Rafael Menezes, PELD/Project—IEAPM)

The Cabo Frio region, specifically, Arraial do Cabo, has a wealth of oceanographic processes at various scales. The topography in the region is characterized by an intense slope that makes the continental shelf relatively narrow. This causes the 100 m isobath to be located 7 km from the coast, exposing its rocky shore. This allows, as previously described, a unique situation, in which there is a proximity between coastal and deep-ocean phenomena. Dominated by the dynamic of BC, which flows southwest, occupying the first 500 m of the water column.

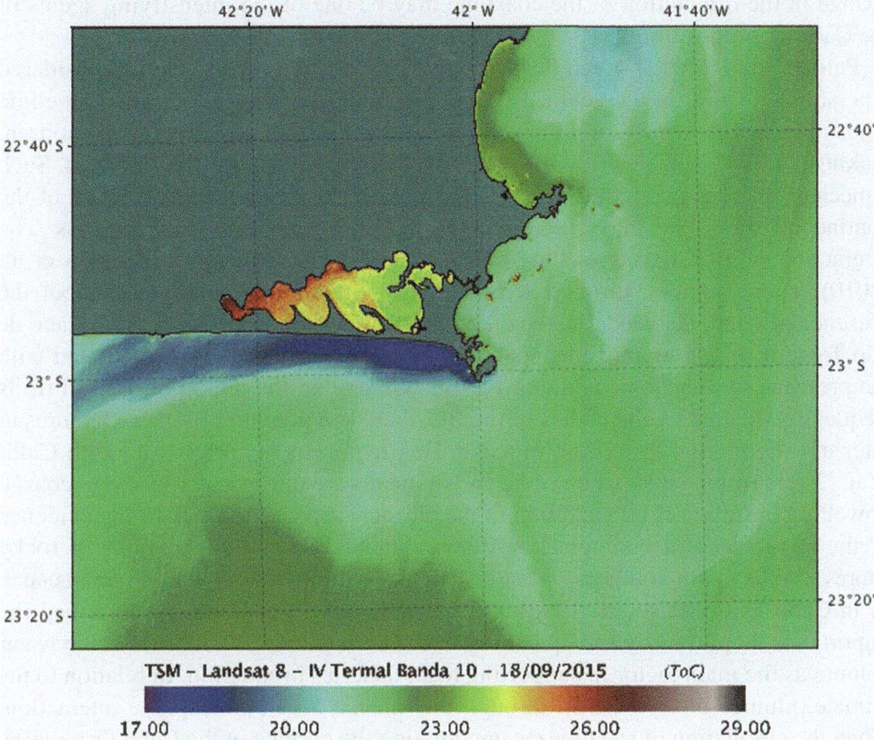

Fig. 1.2 Sea-surface temperature, generated from band 10 of the thermal infrared sensor (TIRS) on board the Landsat-8 satellite, captured on December 19, 2016. In blue, the plume generated by the coldest waters (17–20 °C), generated by the coastal upwelling in CF (Flávio Gomes, IEAPM/UFF)

The Cabo Frio upwelling is defined when the SCAW carried by the BC, at sub-picnoclinic levels (200–500 m), which is transported perpendicularly toward the coast, emerges on the surface, in response to the withdrawal of water caused by E or NE winds parallel to the coast. An intense upwelling of Cabo Frio can cause temperatures below 18 °C, providing a thermal gradient of up to 10 °C between the coast and the oceanic region, typically along the internal border of the BC.

Studies have sought to identify the role of several mechanisms that may explain the greater intensity of Cabo Frio's upwelling when compared to other cases along the Brazilian coast. Rodrigues and Lorenzzetti (2001), for example, explored the role of the topography and geometry of the coastline, which abruptly changes its orientation from N–S to E–W. These authors used a semi-idealized model applied to the Brazilian continental shelf. In an experiment where the idealized ocean with real topography and coast and NE wind of constant magnitude and direction, there was strong evidence of a temperature decline off Cabo Frio, Cabo de São Tomé, and Vitória. This evidence became smaller in Cabo Frio in an additional experiment, where it was considered a smooth and flat bottom, indicating that the substantial change in the orientation of the coastline may be one of the intensifying agents of the Cabo Frio upwelling.

Palóczy et al. (2014) stated that the upwelling of shelf breaks can be considered a persistent feature in the southeastern continental platform, due to the baroclinic adjustment of BC, which forces isotherms to approach the coast close to the bottom, making SACW available closest to the coast for a possible upwelling process. Such a mechanism becomes more expressive in Cabo Frio, given the small width of the continental shelf, and may contribute to intensifying the upwelling process. The interaction of BC with upwelling was explored in more detail by Calado et al. (2010). These authors, through semi-realistic numerical simulations without the presence of wind, detected that the outcrop of cold waters near the coast in Cabo de São Tomé is driven by movements in the bottom and surface layer associated with the presence of a cyclonic eddie of the BC. Given that the region off Cabo Frio is frequently affected by the eddies of the BC, there is a potential for these features to intensify the coastal upwelling process. This hypotesis is corroborate with Callil et al. (2021) recent study when eddie and filaments are intrinsically linked to coastal upwelling in Cabo Frio. In addition to the physical aspects, there is strong evidence of the importance of ecological and environmental impacts, in regions of rocky shores, on the south-southeast coast of Brazil, coming from upwelling points such as in Cabo Frio. Matsuura (1996) and Valentin (2001) had already described the important role played by the upwelling in the enrichment of nutrients in the water column as the main factor in supporting local fisheries production. In relation to the climate, plumes of cold water, through the intense ocean-atmosphere interaction, affect the circulation of sea breezes, modulating the climate in the Cabo Frio region (Franchito et al. 1998).

Wind-Generated Waves

Wind-generated waves are ubiquitous in the ocean, deriving their energy from the wind blowing over water. They are either generated locally or non-locally, many times thousands of kilometers away from the observer. Hence, waves are present even when there is no apparent wind. When growing under the influence of their generating winds, they are called wind-sea. When they are no longer under the influence of the wind, they propagate away from the area and are called swell. When still within the generation area, their growth depends on three components: wind speed, wind duration, and fetch, which is the distance over which the wind blows. Through detailed field measurements—such as the JONSWAP experiment—the relationships between wind speed, wind duration, fetch, and wave growth were parameterized (Hasselmann et al. 1973).

The more important parameters in operational terms are wave height (distance between crest and through), wave length (distance between two successive crests, with its associated wave period), and wave direction (either the direction the waves are coming from or going to, depending on the applied convention). Another important parameter is steepness, the ratio between wave height and wave length and a proxy for its degree of nonlinearity. Theoretically, the simplest approach in wave analysis is to consider a sinusoidal (regular) wave that propagates freely without dissipation. Assuming fundamental principles of fluid dynamics, specifically that the fluid is inviscid, irrotational, and incompressible, a linearized approach is adopted, commonly referred to as Airy Theory or simply Linear Theory.

The basic assumption behind the concept of the wave spectrum is that a complex, irregular, sea state observed in nature does not resemble a sinusoidal wave is a linear superposition of an infinite number of them. Therefore, through mathematical analysis, a complex sea elevation can be decomposed into a summation of regular waves.

The ocean plays a key role as a climate regulator. Ocean currents are very effective redistributing warm, equatorial waters toward higher latitudes. The same applies to colder, polar waters that are carried away toward lower latitudes. Wind-generated waves, on the other hand, are known for their capability to propagate energy, over large distances, with little dissipation. They are less known, however, for being responsible as well for a net mass transport in the wave propagation direction, the Stokes drift (Pereira et al. 2021; Violante-Carvalho et al. 2021). Stokes drift can be thought of as the near-surface ocean current induced from the wave action. The magnitudes of these near-surface currents can be significant. A passive tracer, therefore, will experience a second-order velocity in the wave propagation direction, given by the mean Stokes drift current.

Waves propagate over great circles, without the influence of the Earth's rotation. Even though dissipation is relatively small, wave height is gradually reduced. Because of the turbulent character of the wind, waves are generated within a window of directions around the mean wind direction that leaks out the sides of the fetch. This phenomenon is referred to as directional spreading. Moreover, longer waves with longer periods propagate, in deep water, faster than shorter waves, which are designated as frequency dispersion. Directional spreading and frequency dispersion are responsible for an attenuation of wave height with distance as they move away from the generating storm.

Wave exposure is an important consideration in planning, designing, and operating in the coastal zone. Careful analysis is necessary to determine the main characteristics of the waves such as height, period, and direction of propagation covering intervals of time long enough for the characterization of their spatial and temporal variability. As waves approach the coastline, they begin to interact with the sea bottom going through several transformations in their features. Waves are considered in deep water when the depth is greater than half wavelength. Shallow water waves, on the other hand, are propagating in depths less than one-twentieth of their wavelengths, with a transition zone between these two limits called intermediate water. Similarly as in optical physics, waves experience several changes with the reduction in propagation velocity associated with the reduction in depth. Refraction, to name one of them, changes the wave propagation direction with crests gradually getting parallel with bottom contours, which might increase or decrease wave height depending on the coastline orientation. Around a rocky headland, for instance, refraction will concentrate energy and consequently increase wave height enhancing the potential for erosion.

Other water wave phenomena described by optical physics are diffraction and shoaling. The process of diffraction experienced by water waves is similar to the diffraction of light and sound, with respect to their different wavelengths. In the water, diffraction occurs whenever there is a gradient of wave height along a line of constant phase, such as the crest (Violante-Carvalho et al. 2009). An intuitive and particularly significant example is that of waves encountering an emerged obstacle, such as a breakwater or an island, when they bend and change direction propagating into the sheltered area according to the Huygens-Fresnel principle. Waves decelerate as they propagate into intermediate and shallow water, which causes a reduction in their length while increasing wave height, the so-called shoaling effect.

Wave run-up is defined as the maximum vertical distance above the mean sea level of water excursion along structures. Run-up is an important process, function of the wave steepness and the slope of the rocky shore, causing erosion by direct wave impact or by washing away sediments. Also of relevance is the phenomenon of wave setup, a nonlinear phenomenon causing water level rise due to breaking waves in the surf zone. The elevation in the water level is a fraction (around 10%) of the wave height in the breaker zone, potentially responsible for severe flooding during strong storms.

Undoubtedly, measurements are essential for improving our understanding of several complex physical processes. Satellite oceanography provides a broader picture of the oceans when compared to the few and sparse in situ measurements available. When combined in a complementary sense, in situ and satellite data may generate additional information, especially in regions with lack of in situ observations. More specifically regarding wind-generated waves, satellite-borne altimeters have provided highly accurate measurements of wave heights with extensive temporal and spatial coverage since the early 1990s. Their continuous operation has yielded a substantial and valuable dataset. Gommenginger et al. (2019). Despite its relevance in operational terms, wave height is an integrated parameter not sufficient to fully characterize a sea state.

The synthetic aperture radar (SAR), on the other hand, can yield the full, directional wave spectrum with the known limitations in its high frequency band (Hasselmann et al. 2013). A myriad of satellites, for instance, ERS-1&2, RADARSAT, Envisat, and the currently operational Sentinel1-A&B twins are still continuously collecting nearly 30 years of swell data with tens of millions of observations globally distributed. In their seminal paper, Hasselmann and Hasselmann (1991) presented an analytical relationship for the mapping of a wave spectrum into a SAR image spectrum, the so-called Forward Transform, employing modulation transfer functions (MTFs) with more relevance to hydrodynamic, tilt, and velocity bunching (Hasselmann et al. 1985; Santos et al. 2021). The nonlinear Forward Transform was a major achievement, with its expression later extended to include the SAR look cross spectra (Engen and Johnsen 1995).

The lack of in situ measurements in the Southern Hemisphere is well known, and the South Atlantic is no exception. More specifically, studies based on wave measurements over the Brazilian coast are up to the present spatially sparse, limited mainly to its most important oil fields, located in Campos Basin (Violante-Carvalho et al. 2004; Pereira et al. 2017). Most of the wave buoys under supervision of the National Oceanic and Atmospheric Administration (NOAA) are located in mid-latitudes in the Northern Hemisphere mainly in relatively shallow waters. The other major source of directional wave data is the network deployed along the coasts of Western Europe, as well in comparatively shallow waters. The majority of the Brazilian coast is located in the tropical zone, being strongly affected by swell all year round and therefore with most of the energy in the low frequency portion of the spectrum (Violante-Carvalho et al. 2004). Multi-modal spectra are another common feature in the southernmost part of the Brazilian coast, with an average of only one fourth of the cases classified as unimodal (Violante-Carvalho et al. 2004). In an attempt to overcome the lack of in situ measurements, Brazil's PNBOIA (National Buoys Program) was launched in April 1997 along with the pilot program GOOS (Global Ocean Observing System-Brazil). The goals of this program encompass the collection of metocean data in the Atlantic Ocean along the Brazilian coast by means of a network of moored buoys and drifters supporting meteorological and oceanographic activities important to public institutions and companies (data freely downloadable at http://www.goosbrasil.org).

The shortage of in situ wave data in this region has resulted in numerical modeling becoming the most widely used tool for the investigation of its wave climate (Parise and Farina 2012). Considering 40 years of spectral data from the European Centre for Medium-Range Weather Forecasts employing the ECWAM model, Portilla-Yandun et al. (2016) proposed a methodology for wave spectra partitioning and long-term statistics. A global climatology is available in the GLOSWAC database, presenting a probability distribution of wave spectral partitions (data freely downloadable at http://improlife.xyz/).

Accurate information about wind waves is of importance to a large number of operational, scientific, and recreational activities and several marine processes. Particularly, coastal dynamics is closely linked to the energy released by them in the surf zone. Waves induce turbulence and motion through the ocean up to depths comparable to their wavelengths, affecting the thickness of the mixed layer. At the air-sea interface, the exchange of heat and gases is strongly related to the turbulence induced by them and hence acting as one of the key elements of the global climate balance. Wind waves have fascinated and baffled scientists, poets, and the layman for innumerable generations, both past and future.

References

Andrew NL, Mapstone BD (1987) Sampling and description of spatial pattern in marine ecology. Oceanogr Mar Biol Ann Rev 25:39–90

Calado L, Gangopadhyay A, da Silveira I (2008) Feature-oriented regional modeling and simulations (FORMS) for the western South Atlantic: Southeastern Brazil region. Ocean Model 25:48–64

Calado L, Silveira ICA, Castro BM (2010) Eddy-induced upwelling off Cape São Tomé, 22°S, Brazil. Cont Shelf Res 30:1181–1188

Calado L, Rodríguez OC, Codato G, Xavier FC (2018) Upwelling regime off the Cabo Frio region in Brazil and impact on acoustic propagation. J Acoust Soc Am 143:EL174–EL180. https://doi.org/10.1121/1.5026244. PMID: 29604691

Calil PHR, Suzuki N, Baschek B, da Silveira ICA (2021) Filaments, fronts and eddies in the Cabo Frio coastal upwelling system, Brazil. Fluids 6:54. https://doi.org/10.3390/fluids6020054

Castro BM (1996) Correntes e massas de água da plataforma continental norte de São Paulo. Livre-Docente Thesis, Instituto Oceanográfico da Universidade de São Paulo

Coutinho R (2004) Programa Nacional da Biodiversidade - PRONABIO Projeto de Conservação e Utilização Sustentável da Diversidade Biológica Brasileira – Probio: Sub-Projeto Avaliação e Ações Prioritárias para a Zona Costeira e Marinha Grupo de Ecossistemas: Costões Rochosos. [S.I.]

Engen G, Johnsen H (1995) SAR-ocean wave using image cross spectra. IEEE Trans Geosci Remote Sens 33:1047–1056

Franchito SH, Rao VB, Stech JL, Lorenzzetti JA (1998) The effect of coastal upwelling on the sea breeze circulation at Cabo Frio, Brazil: a numerical experiment. Ann Geophys 16:866–881

Gommenginger C, Chapron B, Hogg A, Buckingham C, Fox-Kemper B, Eriksson L, Soulat F et al (2019) SEASTAR: a mission to study ocean submesoscale dynamics and small-scale atmosphere-ocean processes in coastal, shelf and polar seas. Front Mar Sci 6:457

Hasselmann K, Barnett T, Bouws E, Carlson H, Cartwright D, Enke K, Ewing J, Gienapp H, Hasselmann D, Kruseman P, Meerburg A, Muller P, Olbers D, Richter K, Sell W, Walden H (1973) Measurements of wind-wave growth and swell decay during the Joint North Sea Wave Project (JONSWAP). Deut Hydrogr Z 8:1–95

Hasselmann K, Raney RK, Plant MJ, Alpers W, Shuchman RA, Lyzenga DR, Rufenach CL, Tucker MJ (1985) Theory of synthetic aperture radar ocean imaging: a MARSEN view. J Geophys Res Oceans 90:4659–4686

Hasselmann K, Hasselmann S (1991) On the nonlinear mapping of an ocean wave spectrum into a synthetic aperture radar image spectrum and its inversion. J Geophys Res 96(C6):10713. https://doi.org/10.1029/91JC00302

Hasselmann K, Chapron B, Aouf L, Ardhuin F, Collard F, Engen G, Heimbach P, Janssen P, Johnsen H, Krogstad HE, Lehner S, Li JG, Li X, Rosenthal W, Schulz-Stellenfleth J (2013) The ERS SAR wave mode – a breakthrough in global ocean wave observations. ESA SP 1326:167–197

Lee TN, Pietrafesa LJ (1987) Summer upwelling on the southeastern 224 continental shelf of the U.S.A. during 1981-circulation. Prog Oceanogr 19:267–312

Mascarenhas AS, Miranda LB, Rock NJ (1971) A study of oceanographic conditions in the region of cabo frio. Costlow J D 1(285):308

Matsuura Y (1996) A probable cause of recruitment failure of the Brazilian sardine *Sardinella aurita* population during the 1974/75 spawning season. S Afr J Mar Sci 17:29–35

Palóczy A, da Silveira I, Castro B, Calado L (2014) Coastal upwelling off Cape São Tomé (22°S, Brazil): the supporting role of deep ocean processes. Cont Shelf Res 89:38–50

Pappalardo M, Maggi E, Geppini C, Pannacciulli F (2018) Bioerosive and bioprotective role of barnacles on rocky shores. Sci Total Environ 1:619–620:83–92

Parise C, Farina L (2012) Ocean wave modes in the South Atlantic by a short-scale simulation. Tellus A Dyn Meteorol Oceanogr 64(1):17362

Pereira H, Violante-Carvalho N, Nogueira I, Babanin A, Liu Q, Pinho U, Nascimento F, Parente C (2017) Wave observations from an array of directional buoys over the southern Brazilian coast. Ocean Dyn 67(12):1577–1591. https://doi.org/10.1007/s10236-017-1113-9

Pereira H, Violante-Carvalho N, Fabbri R, Babanin A, Pinho U, Skvortsov A (2021) An algorithm for tracking drifters dispersion induced by wave turbulence using optical cameras. Comput Geosci 148. https://doi.org/10.1016/j.cageo.2020.104654

Portilla-Yandun J, Salazar A, Cavaleri L (2016) Climate patterns derived from ocean wave spectra. Geophys Res Lett 43(22). https://doi.org/10.1002/2016GL071419

Rodrigues RR, Lorenzzetti JA (2001) A numerical study of the effects of bottom topography and coastline geometry on the Southeastern Brazilian coastal upwelling. Cont Shelf Res 21:371–394

Santos F, Santos A, Violante-Carvalho N, Carvalho L, Brasil-Correa Y, Portilla-Yandun J, Romeiser R (2021) A simulator of Synthetic Aperture Radar (SAR) image spectra: the applications on ocean swell waves. Int J Remote Sens 42(8):2981–3001. https://doi.org/10.1080/01431161.2020.1847352

Silveira ICA, Schmidt A, Campos EJD, Godoi SS, Ikeda YA (2000) Corrente do Brasil ao largo da costa leste Brasileira. Rev Bras Oceanogr 48:171–183

Suguio K (2001) Geologia do Quaternário e mudanças ambientais (Passado + Presente = Futuro?). Paulo's Comunicação e Artes Gráficas, São Paulo, p 300

Valentin JL (2001) The Cabo Frio upwelling system, Brazil. In: Seeliger U, Kjerfve B (eds) Coastal marine ecosystems of latin America. Springer, Berlin, pp 97–105

Vilano WF, Souza CRG (2011) Biogeografia de costões rochosos e sua importância para os estudos do Quaternário. XIII ABEQUA

Villwock JA, Lessa GC, Suguio K, Angulo RJ, Dillenburg SR (2005) Geologia e geomorfologia de regiões costeiras. In: Souza CRG, Suguio K, Oliveira AM, Oliveira PE (eds) Quaternário do Brasil. Holos Editora, Cap. 5, pp 94–113

Violante-Carvalho N, Ocampo-Torres F, Robinson I (2004) Buoy observations of the influence of swell on wind waves in the open ocean. Appl Ocean Res 26:49–60

Violante-Carvalho N, Paes-Leme R, Accetta D, Ostritz F (2009) Diffraction and reflection of irregular waves in a harbour employing a spectral model. An Acad Bras Ciênc 81:837–848

Violante-Carvalho N, Skvortsov A, Babanin A, Pereira H, Uggo P, Esperança PTT (2021) The turbulent dispersion of surface drifters by water waves: experimental study. Ocean Dyn. https://doi.org/10.1007/s10236-020-01423-y

Chapter 2
Biogeographic Patterns of Brazilian Rocky Reef Fauna

Cesar A. M. M. Cordeiro, Sergio R. Floeter, and Carlos E. L. Ferreira

Abstract Rocky shores are prominent features of the Brazilian SE-S coast that define the coastal contour in the region and are interspersed by sand beaches. The intertidal and subtidal zones are discussed here as a continuous along the depth gradient, where large rocky formations plunge into the sea and are bordered by a soft sediment basin. The SE-S rocky shore range overlaps climatic and oceanographic domains, with NE currents and winds, and cold fronts coming from S-SW during the dry season. The SE-S shores represent a gradual change in species composition from tropical to subtropical-temperate provinces, where some taxonomic groups peak in species richness along the Rio de Janeiro and São Paulo shores associated with less heterogeneous climatic and oceanographic domains. Fish, mollusk, ascidian, and sponge species richness decrease southward, while corals decrease in subtropical areas mostly because of their clear association with tropical waters. Although some rocky shores are within marine and coastal protected areas, the total area covered and the real enforcement applied to those protected areas are below desirable levels. These habitats are among the most impacted on the Brazilian coast because of their proximity of large cities and should be prioritized in sustainable development initiatives and coastal management planning.

C. A. M. M. Cordeiro (✉)
Laboratório de Ciências Ambientais, Universidade Estadual do Norte Fluminense Darcy Ribeiro, Campos dos Goytacazes, RJ, Brazil
e-mail: cesarcordeiro@uenf.br

S. R. Floeter
Laboratório de Biogeografia e Macroecologia Marinha, Universidade Federal de Santa Catarina, Florianópolis, SC, Brazil

C. E. L. Ferreira
Laboratório de Ecologia e Conservação de Ambientes Recifais, Universidade Federal Fluminense, Niterói, RJ, Brazil

R. Coutinho, R. A. Christofoletti (eds.), *Brazilian Rocky Shores*, Brazilian Marine Biodiversity, https://doi.org/10.1007/978-3-031-67206-4_2

Geographical and Climatic Context

Rocky shores are a conspicuous feature of the landscape on the Brazilian SE-S coast (Ab'Sáber 2000). Much of the rocky shores in this region delimit the coastal contour being interspersed by sand beaches that spread to the infralittoral in the enlarged siliciclastic shelf of this region (Knoppers et al. 1999). In this chapter, rocky shores included intertidal and subtidal zones because, with a few exceptions, both zones form a continuous along the depth gradient. Usually, large rocky formations plunge into the sea and are delimited by a soft sediment basin. Reef formations were considered only as those permanently immersed and detached from the coast. Rocky formations can be found predominantly from Vitória/ES (20.29 S, 40.28 W) to Laguna/SC (28.61 S, 48.82 W), with the more complex and continuous coastal formations located between Mangaratiba/RJ (22.98 S, 44.04 W) and São Sebastião/SP (23.77 S, 45.80 W), and the southernmost coastal crystalline rocky shore is located in Torres/RS (29.36 S, 49.73 W). Coastal rocky formations found on tropical domains from the northeast Brazilian coast are sparse (Coutinho 2004), punctually found from Ceará to Bahia (e.g., Baía de Todos os Santos—Paz et al. 2012; Nunes et al. 2013; Ilhéus and Itacaré—Santos 2020), and will not be treated here. Submerged rocky reefs are found scattered on shallow waters, and also form large patches of mesophotic habitats that are critical for many endemic and threatened species. These reefs are poorly mapped and were also not included in this chapter. Finally, most of the oceanic islands inside Brazilian territorial waters (São Pedro and São Paulo Archipelago, Fernando de Noronha Archipelago, and Trindade Island) are also dominated by rocky formations and will be briefly debated in a separate section in this chapter.

Climatic and oceanographic domains overlap in this SE-S rocky shore range, which means a unique overlap of biodiversity components. The Brazilian Current and NE winds prevail in the SE-S Brazilian shores, and local climate is classified as a tropical monsoon regime with dry "winters" (Alvares et al. 2013). Cold fronts coming from S-SW are especially influential during 'winter" months in the region (Mazzuco et al. 2018). The coastal orientation deflected to the east at this portion of Brazil favors particular conditions that are associated with the shelf geomorphology, eddies formation and wind regimes that build up conditions to upwelling events to occur locally (Calado et al. 2010). The colder waters from the Southern Atlantic Central Water (SACW) mass influences directly the whole shelf and often reaches the surface, especially during September to March at the Cabo Frio/RJ region (Valentin 2001) and at Cabo de Santa Marta/SC (Campos et al. 2013). Upwelling events in Southeastern Brazil influences the local biota by, for instance, affecting larval abundance, recruitment, and productivity as a whole (Mazzuco et al. 2018). Coastal geomorphology and associated oceanographic characteristics create complex gradients of wave exposition, temperature and depth, increasing habitat diversity and harboring species-rich communities (Coutinho 2004; Pardal et al. 2021, 2023).

Biogeographical patterns associated with regional climatic and oceanographic phenomena will be detailed in the next sections.

Biogeographical Patterns

Subtidal Zone

Earlier delimitations of biogeographic regions within the Brazilian coast varied in setting boundaries according to the availability of information for each studied group (see Barroso et al. 2016), but the southeastern coast is usually considered as a transition region of intermediate characteristics from tropical and subtropical domains (Fig. 2.1). Briggs (1974) included the rocky shore domains within the Paulista province (limits between 21° and 30° S) based on endemism values and oceanographic and geological characteristics. Palácio (1982) compiled information about biogeographical patterns of several taxa (e.g., macroalgae, Porifera, Mollusca, Ascidiacea, Echinodermata, and fishes) on the Brazilian coast, and also discussed previous authors that described general patterns of distinct taxa, maintaining the classification proposed by Briggs (1974). Floeter and Soares-Gomes (1999) supported a similar division by Briggs (1974) but set the northern border around Marataízes/ES, although the highest Gastropoda species richness was found in Campos Basin and Cabo Frio, with most of the species showing tropical affinities within the Paulista Province. Later, Martínez and del Río (2002) extended further the province limits considering the whole Brazilian coast as a single province based on fossil records (Miocene) and extant mollusk fauna composition. Lotufo (2002) also defined wider limits to the Paulista Province: from northern Bahia until Santa Catarina based on gastropod species. Although there was a higher similarity in composition between the faunas from São Paulo, Rio de Janeiro, Santa Catarina, and Paraná, when compared to Bahia, the author set the northern limit of the Paulista Province based on the large dissimilarity in species composition when compared to the Brazilian and Patagonic provinces.

Considering algal groups, Horta et al. (2001) indicated the Espírito Santo coast as a transition zone between the Tropical (Piauí to Bahia coasts) and a Warm

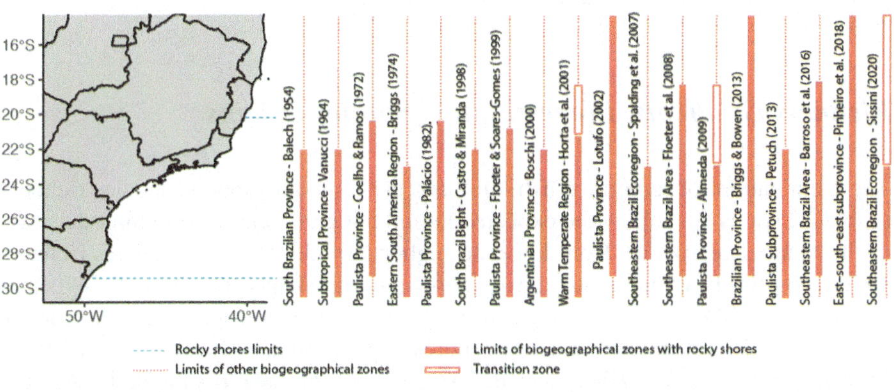

Fig. 2.1 Coastal area dominated by crystalline rocky shores and proposed biogeographic divisions for the Brazilian coast

Temperate Region (Rio de Janeiro to Rio Grande do Sul) due to its richness and mixed composition of species. Recently, the evaluation of coralline red algae richness patterns also indicated a hotspot of species diversity between Bahia and the north of Rio de Janeiro, reinforcing that region as a diversity center and transition zone for coralline red algae on the Brazilian coast (Sissini et al. 2021). Barroso et al. (2016) applied similar boundaries for Gastropods to those indicated by Lotufo (2002) but limiting the northern boundary of the E-SE province to southern Bahia, with the region from Rio de Janeiro to Santa Catarina as one unit and Espírito Santo more similar to NE region. The study conducted by Pinheiro et al. (2018) based on reef fishes found higher similarities between neighboring sites with a closer association between Rio de Janeiro and São Paulo in relation to Santa Catarina and Espírito Santo, the latter being more similar to Bahia. Lotufo (2002) based on the ascidian fauna considered the whole region from northern Bahia to southern Santa Catarina as one unit. It is worth noting that most biogeographic classifications were based on taxonomic groups occupying different habitats, which also usually include nonconsolidated substrates and ecotones (e.g., estuaries). Lotufo (2002), Horta (2000), Floeter et al. (2001, 2008), and Pinheiro et al. (2018) focused their studies on reef-associated species and agreed with other biogeographic studies classifying the rocky reef–dominated region as a transitional and species-rich region on the Brazilian coast.

Despite some disagreement about the northern and southern limits, there is a consensus that throughout the SE-S shores there is a gradual (rather than an abrupt) change in species composition from tropical to subtropical-temperate provinces. Habitat and oceanographic conditions have imposed different filters on marine groups, which directly influenced their geographical distribution limits causing the large frequency of overlap between Espírito Santo and Rio de Janeiro, especially in the north of Rio de Janeiro where upwelling influence is greater (Mazzuco et al. 2015). Some taxonomic groups show a peak in species richness along the Rio de Janeiro and São Paulo shores, reflecting an accumulation of species that gradually decreases to the north and south (Pinheiro et al. 2018) where climatic and oceanographic domains are less heterogeneous (Castro et al. 2006).

Richness Gradients, Processes, and Species Origins

The main pattern observed for fish, mollusk, ascidian, and sponge species richness (Table 2.1) does not follow the expected latitudinal gradient, i.e., a regular decrease in the number of species from lower to higher latitudes. Instead, species richness patterns on the Brazilian coast mostly show a hump-shaped profile with its center located in the SW region over the transition zone between tropical and subtropical regions, which match where rocky shores are dominant (Barroso et al. 2016; Pinheiro et al. 2018; Cruz-Motta et al. 2020). Corals, on the other hand, do decrease in species richness when considering rocky shore domains (Table 2.1), mostly because of their clear association with tropical waters (Leão et al. 2003).

Table 2.1 Species richness of selected taxa along the crystalline rocky shores of southeastern coast and oceanic islands in Brazil

Taxa	ES	RJ	SP	PR	SC	SPSP	FN	TMV
Reef fish								
Chondrichthyes[a]	43	40	46	7	27	23	30	17
Actinopterygii[a]	424	395	427	131	301	202	201	172
Mollusca								
Prosobranchia[b]	288	316	252	154	170	17	126	156
Gastropoda, Bivalvia, Cephalopoda[c, d, e]						48	229	140
Macroalgae[f, g, h]	249	341	258	67	144	36	168	221
Cnidaria								
Scleractinia and Hydrozoa[i, j, k]	13	10	7	25	2	3	11	3
Zoantharia[l, m, n]	9	11	11	3	6		34	37
Porifera[k, o, p, q]	59	102	84	23	45	32	59	28
Ascidiacea[k, r, s, t]	28	32	66	25	34			

ES Espírito Santo, *RJ* Rio de Janeiro, *SP* São Paulo, *PR* Paraná, *SC* Santa Catarina, *SPSP* São Pedro and São Paulo Archipelago, *FN* Fernando de Noronha Archipelago, *TMV* Trindade and Martin Vaz Islands
[a]Pinheiro et al. (2018)
[b]Barroso et al. (2016)
[c]Oliveira et al. (2009)
[d]Gomes et al. (2006)
[e]Gomes et al. (2017)
[f]Horta et al. (2001)
[g]Villaça et al. (2006)
[h]Sissini et al. (2017)
[i]Leão et al. (2016)
[j]Capel et al. (2012)
[k]Amaral et al. (2009)
[l]Bumbeer et al. (2016)
[m]Santos et al. (2016)
[n]Santos et al. (2020)
[o]Muricy et al. (2011)
[p]Moraes et al. (2009)
[q]Moraes and Muricy (2017)
[r] Lotufo (2002)
[s]Rocha et al. (2011)
[t]Rocha et al. (2005)

Jacques Laborel seminal work (1970) (also see Laborel-Degen et al. (2019) for a commented version) described the Abrolhos region as the richest in coral species, decreasing southward, indicating the Cabo Frio region as the limiting zone of distribution for several species of gorgonians and fire corals, and the northern shore of São Paulo as the southern limit of occurrence of scleractinian corals. The steady decline in coral (17 species in Bahia) and hydrocoral species in the granitic shores of SE Brazil is usually associated with the lower water temperature, salinity, and turbidity, as main controls to coral development in higher latitudes (Leão et al. 2016). Further work has shown that Santa Catarina is the true limit for scleractinian coral, with the presence of only *Madracis decactis* and *Astrangia rathbuni*, species known to be resistant to temperature and turbidity variation (Leão et al. 2003; Capel et al. 2012).

Algal groups sum up more than 250 infrageneric taxa in Espírito Santo, but the highest algal richness is found in Rio de Janeiro (341 taxa), decreasing toward southern Brazil (Horta et al. 2001). Fishes account for 515 species (464 Actinopterygii and 81 Chondrichthyes) within the rocky shore domain and show a decrease in species richness southward. However, reef fish assemblages show a high similarity in composition among sites between Bahia and Santa Catarina (Floeter et al. 2001; Pinheiro et al. 2018). Prosobranch mollusks also show higher species richness closer to the northern limits of Brazilian rocky shores with 359 species recorded from south of Bahia to Rio de Janeiro, decreasing to 170 species at the southern limit of rocky shores in Santa Catarina (Barroso et al. 2016).

Following the hypothesis of latitudinal gradients associated with decreasing water temperatures (Floeter et al. 2004, 2005; Brayard et al. 2005), the northern limit of the Brazilian rocky shores is coincident with higher species richness and higher mean temperature, despite the cold-water intrusions (Campos et al. 2000). Also, the southern limit has lower species richness (Horta et al. 2001; Floeter et al. 2008) and lower (and more variable) water temperatures (Campos et al. 1999). Such impoverished fauna at the southern sites of the Brazilian coast is usually attributed to the diminishing numbers of reef-associated species with tropical affinity. Most of these tropical lineages have evolutionary origins in the Caribbean (e.g., sea urchins—Lessios et al. (1999) and McCartney et al. (2000); corals—Nunes et al. (2011), and other groups already cited here). For reef fishes, most species have relatively recent Caribbean origins (Floeter et al. 2008), but for some taxa (e.g., some Blenniidae, Sparidae, Serranidae) other routes of species' origins are also known (Araújo et al. 2020). However, some taxa do not follow this trend such as brown algae (Horta et al. 2001), corals (Fukami et al. 2004), and sponges (van Soest and Hajdu 1997), which have phylogenetic histories prior to the Tethys closure, probably occupying the Southern Atlantic region before the connection with the Caribbean. This pattern indicates a southern route between Atlantic and Indo-Pacific (Horta et al. 2001), but also exporting species to the Caribbean region (Rocha et al. 2008).

The transitional role of the rocky reefs in the Brazilian coast is also supported by the low connectivity separating many populations from tropical to subtropical waters, including genetic evidence for corals (*Siderastrea stellata*—Neves et al. 2007; *Mussismilia hispida*—Picciani et al. 2016; Peluso et al. 2018) and reef fishes (Santos et al. 2006; Affonso and Galetti 2007; Machado et al. 2017). The Espírito Santo coast is evidently a transitional zone within the Brazilian Province, bearing a mix of habitats with rocky shores, rhodolite beds, arenitic and algal reefs overlapping, which slowly changes to rocky shore dominance in the central-southern coasts, roughly south of the Doce River (19° S) (Moura et al. 1999; Mazzuco et al. 2020). Temperature differences can act as barriers to species, as long as differences between close regions exceed the natural variability within them because of physiological constraints of species (Kocsis et al. 2018). Thus, summed to the change in habitat homogeneity, the upwelling area at São Tomé (Palóczy et al. 2014) and Cabo

Frio (Valentin 2001) together with coastal circulation patterns, create filters to species dispersion and establishment further south (Peluso et al. 2018). Turbidity was also indicated as an importance factor associated with changes in benthic cover composition along the Brazilian coast, where southeast Brazil rocky reefs were included as turbid, and the oceanic islands as clear water reefs (Santana et al. 2023). This composition split was also shown by multiple taxa, including macroalgae, corals, crustaceans, mollusks, elasmobranchs, and reef fish, indicating that those subprovinces (southeastern and south) are consistently different from the rest of the Brazilian coast (Cord et al. 2022).

At the community level, few studies have evaluated changes in composition along Brazilian rocky shores (Floeter et al. 2001; Ferreira et al. 2004; Anderson et al. 2020a). Morais et al. (2017) and Aued et al. (2018) analyzed patterns of fish assemblages and benthic communities, revealing very similar patterns to those indicated previously, where rocky shore assemblages and communities presented homogeneity and larger within-site similarity when compared with tropical areas along the coast. Macroecological analyses of reef fishes revealed that they were more influenced by regional and local factors than large-scale processes, especially in rocky reefs (Morais et al. 2017). Bender et al. (2013) indicated the role of habitat differences (i.e., rocky reefs) in community assembly for the southwestern Atlantic, where either functional or taxonomic attributes applied to analyses indicated a high similarity in composition of reef fish communities in southeastern areas of the Brazilian coast, reinforcing the clustering of areas between Espírito Santo and Santa Catarina. A similar pattern in clustering of rocky reefs was found for benthic communities with large dominance of algal groups and a gradual decrease in coral species richness toward southern shores on the Brazilian coast (Aued et al. 2018). The local analysis also indicated a steady decline in fish species at higher latitudes (Pinheiro et al. 2018) and a turnover in composition to subtropical invertebrate species (Lotufo 2002; Barroso et al. 2016). Some subtropical to temperate reef fish species can also be found at lower latitudes but related to deep reefs or upwelling zones (e.g., *Pinguipes brasilianus, Dules auriga*—Anderson et al. 2015; *Serranus aliceae*—Carvalho-Filho and Ferreira 2013), associated with intrusion of the SACW. Finally, Luza et al. (2023) evaluated the functional patterns of reef fish, coral, and algae, and they observed that sea surface temperature, species richness, and region were the main factors determining spatially congruent functional diversity patterns across Brazilian reefs, irrespective of their evolutionary histories. Main differences between southern rocky reefs, northern reefs, and oceanic islands indicated a lower reef fish functional richness, but higher algal functional richness along southern rocky reefs (Luza et al. 2023), which follows the same patterns indicated for species richness along this chapter.

Human-induced biological invasions are also a key process gradually shaping the composition of reef communities, especially in regions with intense ship traffic such as the Santos and Guanabara bays. Fouling and ballast water serve as long-lasting occurring vectors (Carlton 1996; de Messano et al. 2019; Oricchio et al. 2019). Harbor areas on the Brazilian coast, especially in the southwestern region, are typically alien species–enriched habitats (Teixeira and Creed 2020). Consequently, non-native macroalgae, ascidians, corals, and soft corals have been found with more frequency, many being reported as invasive (Ferreira et al. 2006; Lages et al. 2006; Silva et al. 2014; Menezes et al. 2022). However, negative effects of alien species over the benthic elements were detected only for invasive corals (e.g., *Chromonephthea braziliensis* vs. *Phyllogorgida dilatata*—Lages et al. (2006); *Tubastraea* spp. vs. *Mussismiia hispida*—Creed (2006); *Tubastraea* spp. and vs. *Palythoa caribaeorum*—Guilhem et al. (2020); turf-forming algae following vs. *Latissimia ningalooensis*—Carvalho-Junior et al. (2023)), while a large-scale view of impacts over benthic communities is still to be revealed.

Natural invasions are much more difficult to register and a few have been indicated; for instance, the population expansion of the non-native damselfish *Chromis limbata* (Anderson et al. 2020b) has been increasing in densities, although non-negative impacts have been reported. A potential harmful threat to southern rocky reef communities is the lionfish *Pterois* spp. that was observed in Brazilian waters in 2015 which was probably a released vector from aquarium trade (Ferreira et al. 2015). The invasion initiated only in 2018 in Fernando de Noronha and North of Brazilian coast (Luiz et al. 2021). Then, lion fish populations have expanded their ranges along northeastern reefs (Soares et al. 2022), and it is just a matter of time for they reach the rocky reefs with potentially profound impacts (Bumbeer et al. 2018).

Intertidal Zone

Intertidal areas have usually lower species richness compared to subtidal habitats in the same region because of the wide range of environmental stressors (e.g., hydrodynamics, temperature, and UV exposure) acting over the biota living between tides (Menge 2002; Twist et al. 2020). Those stressors act at different levels across the vertical relief creating zones dominated by species best adapted to such levels, and this marked vertical distribution has been debated as a global pattern irrespective of the marine region (Schiel 2004). The vertical zonation is also considered a true pattern for Brazilian intertidal rocky shores (Coutinho 1995), and, such changes in structuring species can also influence richness patterns in rocky shores where there is a negative correlation between tidal height and species richness (Coutinho 2004), i.e., the closer to the water, higher the richness.

Larger scale analysis of alpha and beta diversity of the intertidal communities indicated the Cabo Frio region as an inflexion point for both indicators with decreasing values toward higher latitudes on the Brazilian coast (Cruz-Motta et al. 2020), as observed for subtidal groups in the previous section. Beta diversity was predominantly associated with species turnover at the local scale, indicating small-scale variations as general drivers of assemblages' differentiation (Cruz-Motta et al. 2020). Also, barriers for species dispersion are congruent with those observed for subtidal taxa (Floeter et al. 2008), including the Rio Doce/ES at the north, and the Plata River-Lagoa dos Patos complex at the south, both with generous freshwater inputs to coastal waters that might work as filters for stenohaline species. Another crucial limiting factor to intertidal communities is the substrate availability, which is scarce at the southern border in Santa Catarina (Coutinho 2004). On the other

Table 2.2 Number of species per Phyla, according to state provinces and tidal height, found on intertidal rocky shores reported by Cruz-Motta et al. (2020) supplemented with species registered by Pardal et al. (2023)

	Espírito Santo			Rio de Janeiro			São Paulo			Paraná			Santa Catarina		
Taxonomic group	HT	MT	LT	HT	MT	LT	HT	MT	LT	HT	MT	LT	HT	MT	LT
Cyanobacteria	1	–	–	–	1	1	–	1	1	–	–	–	–	–	13
Chlorophyta	4	–	19	–	3	5	2	9	12	2	–	9	9	–	6
Ochrophyta	4	–	12	–	2	4	–	3	8	–	–	12	2	–	23
Rhodophyta	7	–	53	–	7	10	4	15	28	6	–	31	5	–	–
Tracheophyta	–	–	1	–	–	–	–	–	–	–	–	–	–	–	1
Porifera	–	–	2	–	–	1	–	1	2	–	–	–	–	–	2
Cnidaria	1	–	3	–	–	1	–	–	2	–	–	–	–	–	5
Mollusca	10	–	7	2	10	9	7	18	19	3	–	6	6	–	1
Annelida	1	–	1	–	1	1	–	1	1	–	–	–	1	–	3
Arthropoda	2	–	1	1	2	3	2	4	4	1	–	3	2	–	–
Echinodermata	1	–	2	–	–	2	–	1	1	–	–	2	–	–	–
Bryozoa	–	–	–	–	–	–	–	–	1	–	–	–	–	–	1
Chordata	–	–	–	–	–	1	–	1	1	–	–	–	–	–	–
Total	31	–	101	3	26	38	15	54	80	12	–	63	25	–	55

HT high-tide, *MT* mid-tide, *LT* low-tide

hand, many intertidal species can tolerate a wide range of environmental changes, making those species widely dispersed beyond the limits of the crystalline shores of the S-SE Brazilian coast, ranging up to Maranhão, and reappearing in the south of USA (Klôh et al. 2013; Nunes et al. 2016; Wares 2019).

Several species found on intertidal rocky environments have a wide distribution, such as Chthamalidae (Wares 2019) and Tetraclidae barnacles, besides Mytilidae mussels (Trovant et al. 2013), which are the structuring species of Brazilian rocky shores (Coutinho 1995). Because of the high dominance of few species and wide distribution, there is a large uniformity in the composition of intertidal communities of the Brazilian Province (Cruz-Motta et al. 2020). The next section deals with richness patterns of the dominant and most common species in the intertidal zone. However, as indicated above, most of these species are widespread along SE rocky shores, and richness patterns are more variable locally than between subregions (Pardal et al. 2023) (Table 2.2).

Richness Gradients, Processes, and Species Origins

General richness of species in the intertidal zone indicated a higher number of species in São Paulo except for macroalgae that were more diverse in Espírito Santo shores, as observed for the same group in the subtidal zone. No clear pattern can be observed considering tide height maybe because of lack of data from mid-tide in Cruz-Motta et al. (2020), but richness of species was always higher at the low-tide zone, as expected (Coutinho 1995). Nine species of barnacles are found on the Brazilian coast, and two native species (*Chthamalus bisinuatus* and *Tetraclita stalactifera*) dominate the upper and middle intertidal zones all over the south and southeastern Brazilian shores (Coutinho 2004). The highest richness of barnacle species ($n = 5$) was found in Paraná, at Ilha do Mel, close to Port of Paranaguá, but with a large proportion of alien species ($n = 4$) (Klôh et al. 2013). In the upper intertidal zone, *Chthamalus bisinuatus* can be found in sympatry with the native *Chthamalus proteus*. Both barnacle species are externally indistinguishable (Masi et al. 2009), but *C. proteus* is more associated with euryhaline environments, such as estuaries (Young 1993). Despite not being dominant, nonindigenous species are common elements on Brazilian rocky shores, especially next to harbors. The mobile fauna found at the upper intertidal zone is also poor in species number, where *Echinolittorina lineolata* is the most abundant species, also slightly overlapping in vertical occupation with *Littorina flava* (Rios 1994). Highly mobile isopods (*Lygia exotica*) also live associated with crevices in that zone but foraging all along the intertidal region (Christofoletti et al. 2011). The green crab *Pachygrapsus transversus* is also found occupying the intertidal zone, usually in larger densities at the mid-intertidal region (Flores & Negreiros-Fransozo 1999). These common mobile invertebrates from the upper intertidal are grazers and constitute a crucial part of trophic control of the few algal species (e.g., *Centroceras*, *Enteromorpha*, *Ulva*, and *Cladophora*) occupying this region susceptible to larger periods of air exposition (Coutinho 2004, Christofoletti et al. 2010).

The species *Mytilaster solisianus* can be dominant in the upper portion of the mid-intertidal rocky zones in some sites, and *Brachidontes darwinianus* may occur in sympatry in estuaries (Trovant et al. 2016), following a similar trend of the Chthamalid species cited above. However, at a few shores, *M. solisianus* is absent from the middle intertidal zone, with examples from north of São Paulo and south of Rio de Janeiro shores influencing negatively local richness patterns (Pardal et al. 2023). The absence or low dominance of bivalves in Brazilian rocky shores, in opposition to what is observed worldwide (Schiel 2004; Murray et al. 2006), is mainly associated with recruitment patterns (Mazzuco et al. 2015, 2018) and herbivory pressure (Christofoletti et al. 2011). Still, the mid-intertidal zone is mostly dominated by *Tetraclita stalactifera*, often sharing space with other less frequent barnacle species (Klôh et al. 2013), and some previous (*Perna perna*—Souza et al. 2010) and recent invasive bivalves (*Isognomon bicolor*—López et al. 2014). Although considered as a nonindigenous species, *P. perna* has been established in the Brazilian Province for 2000 years (Souza et al. 2010; Calazans et al. 2021), and developed mussel beds can increase local structural complexity and aggregate considerable diversity of encrusting and vagile species (Jacobi 1987; Borthagaray and Carranza 2007). Algal groups are more diverse in this zone in comparison to the upper intertidal, with the genera *Chaetomorpha*, *Dictyota*, *Gracilaria*, *Acanthophora*, *Hypnea*, *Jania-Amphiroa* complex and other canopy-forming species being common (Brasileiro et al. 2009; Faveri et al. 2020). Mobile fauna is also more diverse in this zone, with grazers dominated by limpets (*Lottia subrugosa*, *Fissurella rosea*, and *Fissurella clenchi*) (Magalhãens and Coutinho 1995), as well as secondary consumers such as whelks (*Stramonita brasiliensis*, *Leucozonia nassa*, and *Claremontiella nodulosa*) and crabs (*Menippe nodifrons* and *Eriphia gonagra*).

At the subtidal fringe, frondose macroalgae beds (e.g., *Sargassum* spp.) and sabellariid reefs (*Phragmatopoma caudata*) can congregate a substantial number of associated taxa. Crustacean and Gastropoda species are the most studied group associated with algal fronds in the intertidal, with richness of Decapoda and Peracarida summing up to 23 (Széchy et al. 2001; Barros-Alves et al. 2017), and 42 (Jacobucci et al. 2018) species, respectively, and, Gastropoda groups ranging from 14 to 68 species (Longo et al. 2019; Colares et al. 2021). Sabellariid reefs can be found throughout the Brazilian coast but are scattered along SW rocky shores, depending on specific conditions of wave exposure and sedimentation (Nunes et al. 2016). Several groups can be found in association with these gregarious polychaetes, such as Decapoda crustaceans (five species in Micheletti-Flores and Negreiros-Fransozo (1999), and eight species in Bosa and Masunari (2002)). Unfortunately, the available information showed great variation in sampling effort and methodology, which is insufficient to depict general patterns about associated fauna with either macroalgae, mussel beds, or sabellariid reefs.

Looking at evolutionary time, processes influencing the biogeographic past of intertidal species seem to be coherent with those observed for subtidal species. The phylogeography and distributional patterns of intertidal organisms are congruent with those observed for subtidal species, where the principal barriers found along the Atlantic helped shape the actual diversity and endemicity patterns. Considering the scale of dispersal as the main vicariant process, the width of the Atlantic Ocean, the Amazon outflow, and the closing of the Isthmus of Panama were the three major barriers that have contributed to the speciation of many species in the western Atlantic (Rocha 2003; Floeter et al. 2008; Reid 2009; Pinheiro et al. 2018), including those occupying the intertidal habitats. Despite the wide distribution along the Brazilian coast, some of the dominant and structuring species in the intertidal are endemic to the Southwestern Atlantic and have a recent past. For instance, the scorched mussel *M. solisianus* also appears to have an ancestor dispersed from the Caribbean region by crossing the Amazon filter, ca. 2.6 Ma (Trovant et al. 2016), and the same is hypothesized for *P. transversus* (Schubart et al. 2005) despite occurring on both sides of the Atlantic. The *E. lineolata*, endemic to the Brazilian province, seems to have originated in the Thethyan period, but its basal position makes speculation about the taxon's origin difficult. The same is not true for the oceanic island endemic *Echinolittorina vermeiji* found only in Brazilian oceanic islands, which have a Caribbean sister species, and apparently diverge from it after the Amazon drainage established in the Atlantic in the Late Miocene (Reid 2009).

Recent patterns of communities' composition are usually reported for restricted scales on Brazilian rocky shores (Paula et al. 2020). However, based on the surveying of 62 rocky shores ranging from north of Rio de Janeiro (Armação dos Búzios/RJ) to south of São Paulo (Itanhaém/SP), Pardal et al. (2023) and Cordeiro et al. (2024) sampled the relative cover of sessile taxa and density of mobile invertebrates in a large gradient encompassing >50% of the rocky shore domains in SE Brazilian coast. The analysis of the presence and absence of species along sampled shores

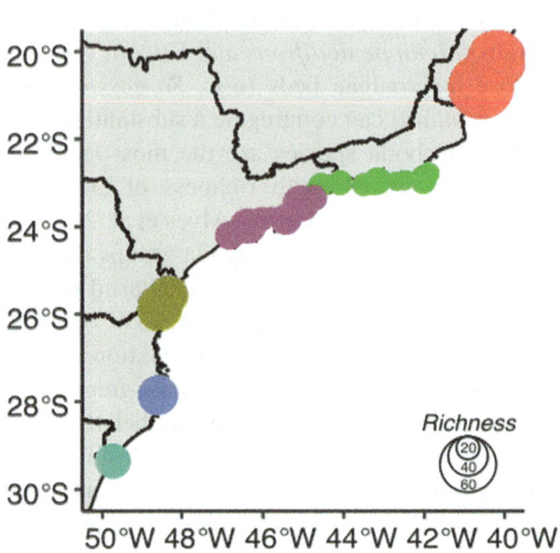

Fig. 2.2 Patterns of similarity and species richness of intertidal communities based on Cruz-Motta et al. (2020) and Pardal et al. (2023). The hierarchical clustering was based on Jaccard coefficient applied to sites ($n = 83$) community species' composition (presence and absence) using the average linkage method

indicated higher similarity between close sites and no evidence of richness gradients from north to south. Clusters obtained from a classification of environmental and biotic indicators (sea surface temperature [SST], wave exposure, average depth, and chlorophyll-a [Chl-a]) for the studied region also support higher similarities among neighbor shores (Fig. 2.2). Sites at north of Guanabara Bay (RJ) were clustered together and associated with higher Chl-a concentration and low SST, indicating effects of the upwelling (Coelho-Souza et al. 2012). Sites south from Rio de Janeiro and north from São Paulo (from Mangaratiba/RJ to São Sebastião/SP) were similar especially because of their variable wave exposure associated with the complex coastal contour forming small inlets either facing or protected from prevailing swells (Pardal et al. 2021). Finally, sites close to Santos and Guanabara bays had large similarities concerning Chl-a concentration, which is associated with the large productivity of those eutrophic populated regions (Miyashita et al. 2009; Villac and Tenenbaum 2010). The spatial dependency of sampled sites indicates that local factors may overcome the influence of regional processes at the scale investigated. The general forcing factors of recent patterns and the local-scale influence cited here are just to better frame dominant processes acting on rocky shore communities but will be better developed and discussed in the following publications (Pardal et al. 2021, 2023; Cordeiro et al. 2024).

Benthic-pelagic coupling is especially critical for intertidal taxa, while oceanographic phenomena modulate recruitment and settlement patterns of structuring species (Menge 1991; Connolly and Roughgarden 1998), helping to shape communities and regulating richness patterns (Burrows et al. 2008). Oceanographic and meteorological conditions on the SE coast of Brazil were described in another chapter of this book and will be only briefly discussed here. As indicated previously, currents, eddies, and the coastal morphology have a greater influence on regional processes shaping communities in Brazilian rocky shores. In the SE-S coast, the effects of the cold-enriched waters from SACW are spread throughout but are more intense in northern Rio de Janeiro (Fig. 2.2) because of Cabo de São Tomé and Cabo Frio upwellings. These nutrients boost local productivity, enhancing phytoplankton growth and the productivity of communities on these rocky shores (Guimaraens and Coutinho 1996; Coelho-Souza et al. 2012), but in the present evaluation, there were no elements to directly link productivity to species richness.

Cold front events associated with winds from the south quadrant, with a higher frequency during the autumn and winter months, affect many of these rocky shores. This wind regime disables the coast upwelling system, being more intense on southernmost sites (Campos et al. 2013). Despite influencing the regional scale, the effects of cold fronts on structuring benthic communities on south-southeastern Brazilian rocky shores have yet to be investigated. Mazzuco et al. (2018) evaluated the impact of cold fronts on the settlement and larvae abundance in areas between Arraial do Cabo/RJ and São Sebastião/SP and observed a significant influence of the seasonal incursion of SW-S-SE winds altering coastal larval transport and affecting positively barnacles and mussels, which are foundation species of intertidal habitats (Coutinho 1995).

Regional-scale processes, such as ocean productivity and water temperature, are determinant to rocky shore communities in Brazil (Cruz-Motta et al. 2020) and elsewhere (e.g., Bosnan et al. 1987; Blanchette et al. 2006; Lathlean et al. 2015). However, local processes (e.g., shore structural complexity, wave exposure, and community composition) seem to be crucial to understand intertidal communities on Brazilian rocky shores (Flores et al. 2015). The evidence and data debated herein indicated an effect of local-scale features on richness patterns additionally to regional tendencies associated with the SACW influence. Low levels of wave exposure were related to higher densities of barnacles than bivalves (Christofoletti et al. 2011), and such determination of the foundation species possibly affects general patterns of species richness. At reduced wave exposure, *M. solisianus* may be absent (Christofoletti et al. 2011), being substituted by oysters in the middle intertidal zones or suppressed by competing species as *C. bisinuatus* in high wave exposure shores. On the other hand, the barnacle *T. stalactifera* may be replaced by *P. perna* mussels on highly exposed shores (Coutinho 2004).

As general patterns, rocky shores from the Cabo Frio region are more similar considering biological (community composition and Chl-a) and environmental variables in relation to other rocky shores. Rocky shores close to large bays (Santos and Guanabara) had similar communities and high Chl-a values associated with eutrophic conditions, while all other rocky shores formed a gradient associated with different wave exposure degree and water temperature (Fig. 2.3). All observations reinforce the importance of the oceanographic context shaping intertidal communities (Menge et al. 2003), with a regional gradient in SST associated with the Cabo Frio upwelling, besides wave exposition and productivity (Chl-a) influencing rocky

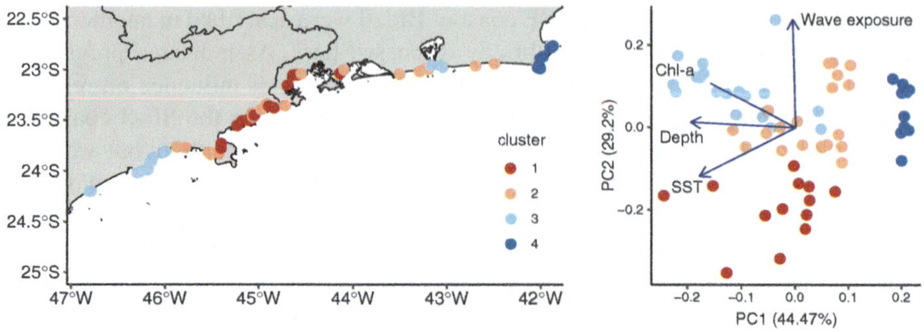

Fig. 2.3 Rocky shores sampled ($n = 62$) for intertidal community composition along the Southeast Brazilian coast by Cordeiro et al. (2024). Clusters were calculated using k-means over environmental data similarity matrix (Euclidean distance). PCA was made based on standardized values (i.e., mean equals to zero and variance to unit)

intertidal communities in the SE Brazilian coast. However, top-down processes acting in local scale may be determinant to species composition (O'Connor et al. 2011) on other population-level parameters (e.g., biomass and productivity) not discussed here.

Offshore Rocky Shores

The Brazilian oceanic islands, namely, Fernando de Noronha Archipelago (FN), São Pedro and São Paulo Archipelago (SPSP), and Trindade and Martin Vaz Islands (TMV), are true rocky formations but differ from coastal rocky shores in their climatic domain, which are all located on tropical waters. Atol das Rocas is one of the Brazilian oceanic systems, but it consists of a biogenic reef consolidated by calcareous algae and vermetids (Soares et al. 2011) and will not be included in further discussions because of the great difference in its reef formation. FN and TMV have a volcanic origin and are composed of steep shores and rocky terraces (Castro 2009), while SPSP is a tiny (~130 m^2) emerged tip of the Meso-Atlantic Ridge with steep shores and limited shallow areas (Campos et al. 2009).

Oceanic islands are unique biogeographic outposts, in general showing low species richness and higher endemicity levels when compared to continental domains because of area and connectivity limitations (MacArthur and Wilson 1967). The Brazilian oceanic islands are no exception for these patterns, and several taxonomic groups have low species richness and several endemics (Hachich et al. 2015; Quimbayo et al. 2018). In general, each island has distinct faunal components and may be defined as independent subprovinces due to high endemism (Spalding et al. 2007; Floeter et al. 2008; Pinheiro et al. 2018). The endemism for reef fish species is below 10% (SPSP = 9%, FN = 4.5% and TMV = 6%—Pinheiro et al. 2018), but it goes beyond this level when adding other taxa, which conferred the status of separate ecoregions in Spalding et al. (2007). Despite their singularities, some islands present higher similarity with continental areas, as in Barroso et al. (2016), where FN and TMV were close together near continental sites, while SPSP formed an isolated branch because of its low species richness (n = 17) and restricted reef area. The TMV have higher connectivity with continental areas because of submarine mounts that act as stepping stones for species (Hachich et al. 2015; Pinheiro et al. 2017), while populations from SPSP and FN may rely mostly on larvae brought by currents. This argument is reinforced by the absence of long route larvae transportation for the TMV islands (Endo et al. 2019), which implies to some local populations (Sally Lightfoot crab *Grapsus grapsus*) to rely on self-recruitment (Freire et al. 2021). Curiously, population connectivity among coral species and crustaceans in the oceanic islands indicates gene exchanges among island populations (Teschima et al. 2016; Peluso et al. 2018; Freire et al. 2021) leaving open questions about their level of isolation. Despite their isolation, island communities are still more similar between them and share more species with the Brazilian coast than with other Atlantic oceanic islands (Floeter et al. 2001; Rocha 2003). However, the presence of Paleogene and Neogene species living together, as well as more recent species in the Brazilian oceanic islands, indicates that they can serve as evolutionary hotspots in the South Atlantic (Pinheiro et al. 2017, 2018).

At the community scale, all islands showed similar patterns of benthic composition dominated by primary producers (turf and macroalgae) and with low complexity components (e.g., zoanthids and massive corals—Aued et al. 2018). Reef fish assemblages are dominated by planktivores and omnivorous species (Morais et al. 2017), which is compatible with plankton-based trophic chains (Ferreira et al. 2004). Also, the density and biomass of fish species tend to be higher when compared to continental reefs, especially because of the contribution of pelagic species and lesser influence from fishing (Morais et al. 2017).

Data related to intertidal environments in the Brazilian oceanic islands is scarce, despite the easy access to the intertidal zone in most of them (Andrades et al. 2018). However, high swells and steep areas are additional risks to intertidal work in oceanic islands and may restrict research possibilities to some sites and certain seasons in each island. Despite the steepness of those rocky shores, tide pools are abundant on intertidal flats of some areas in all Brazilian oceanic islands and harbor a high number of endemic species, even higher than found for the same taxa in the subtidal region (Andrades et al. 2017). For instance, the endemism level of tidepool-resident fish species can reach 44% at FN (n = 8 species) and 38% at Trindade Island (n = 5), being dominated by herbivorous species that structure local assemblages (Andrades et al. 2018). In contrast, sessile and mobile organisms living on zones subject to air exposition during low tides show limited species richness, and some important groups can be monospecific in each island, such as limpets and littorinids (Eston et al. 1986; Simone 2008; Barroso et al. 2016). Encrusting calcareous algae are abundant in intertidal rocky shores with high inclination and walls of tide pools in TMV and SPSP, which may indicate a large grazing pressure (pers. obs.). Filter feeding species are rare as a probable reflex of the oligotrophic waters, with small mitylid patches found at FN (Eston et al. 1986, pers. obs.). It is also worth noting that the most conspicuous intertidal taxa, i.e., barnacles, are virtually absent from the Brazilian oceanic islands, being found only at Trindade Island in very low densities (*Chthamalus* sp.—pers. obs.) and rarely in FN (*Megabalanus coccopoma*—Eston et al. 1986). Finally, the yellow crab *Grapsus grapsus* as a potential key species is worth mentioning. The yellow crab is a conspicuous and important component for intertidal communities of all Brazilian oceanic islands and a voracious omnivorous predator, which can feed from algae to conspecifics, imposing a high predation pressure to intertidal species (Freire et al. 2011).

Final Remarks

Our understanding of ecological patterns of the communities inhabiting the rocky shores of the southeastern-south of Brazil indicated ecological drivers as more important than historical ones (Lotufo 2002), since there are no evident biogeographical barriers or filters within the region occupied by coastal rocky shores. Despite the upwelling area located in the Cabo Frio region could represent the geographical limit for some species, it acts as a turning point for most species by aggregating species from both tropical and subtropical-temperate domains. Warm temperate affinity reef fishes that are widely distributed along the subtropical shallow habitats of rocky reefs domain, where ACAS and BC merge, have their distribution limited to the north until the northeastern coast in some habitats of Bahia state (Carvalho-Filho 1999; Pinheiro et al. 2018). However, some temperate affinity fishes that inhabit mesophotic reefs along the SE-S coast [e.g. *Pronotogrammus martinicensis* (Carvalho et al. 2009) and *Halichoeres sazimai* (Luiz et al. 2009)] can be detected in shallow reefs in the Cabo Frio region, where upwelling processes take place. Thus, oceanographic and habitat characteristics overlap around the rocky shores of the Cabo Frio region, especially at Arraial do Cabo, but those limits are the confluence of wider environmental gradients. Still, habitat availability is limited for species associated with consolidated substrate with large gaps between stable environments: around 120 km of a long sand stretch between Santa Marta Cape/SC and Torres/RS, and >600 km between Torres/RS and La Coronilla (Uruguay). However, artificial structures, such as the breakwaters of Rio Grande/RS, may act as stepping stones (Steigleder et al. 2019) for species dispersion into new habitats on the Argentinean coast.

The number of intertidal species did not show the same decrease observed for subtidal taxa towards southern rocky shores. Indeed, for all rocky shore/reef-associated groups there is a sign of main oceanographic characteristics determining local communities' composition within the crystalline shores' domains. Sea surface temperature and productivity are the most prominent environmental factors acting at regional scale, but the similarity of spatially closest shores indicates the importance of local-scale processes, which may include biotic interactions. In order to confirm such larger influence of abiotic drivers, large experimental approach and long-term monitoring focused on stratified sampling along natural gradients are potential next steps in rocky shore research.

In this chapter, contemporaneous pressures derived from anthropogenic causes were not discussed as drivers of species richness patterns. Ocean warming may also pose additional challenges to conservation actions along Brazilian subtropical shores. Tropicalization examples in the Southwestern Atlantic are rare (e.g., Araújo et al. 2018; Franco et al. 2020) and yet to be reported for rocky shores, but deserve deeper investigation with special attention to species occupying a wide depth range (Coelho et al. 2021) and to possible buffer effects of the upwelling on temperature rise (Inagaki et al. 2020). Finally, as observed for inter and subtidal taxa, the crystalline shores of the Brazilian coast are a hotspot of biodiversity and happen to be located on the most populated part of Brazil. Although some rocky shores are within marine and coastal protected areas, the total area covered and the real enforcement applied to those protected areas are still below desirable levels (Magris et al. 2020). Thus, these rich habitats are among the most impacted on the Brazilian coast because of the proximity of large cities (Lotufo 2002; Ferreira and Rosso 2009) and should be prioritized on sustainable development initiatives and coastal management planning.

Acknowledgments CAMMC is grateful to Fundação de Amparo à Pesquisa do Estado do Rio de Janeiro (E-26/202.310/2019) for the post-doctoral fellowship. CELF and SRF are grateful for continued CNPq grants over the years. CELF and CAMM are also supported by FAPERJ grants.

References

Ab'Sáber AN (2000) Fundamentos da Geomorfologia Costeira do Brasil Inter e Subtropical. Rev Bras Geomorf 1(1):27–43

Affonso P, Galetti PG Jr (2007) Genetic diversity of three ornamental reef fishes (families Pomacanthidae and Chaetodontidae) from the Brazilian coast. Braz J Biol 67:925–933

Alvares CA, Stape JL, Sentelhas PC, Gonçalves JLDM, Sparovek G (2013) Köppen's climate classification map for Brazil. Meteorol Z 22:711–728

Amaral FMD, Ramos CAC, Leão ZMAN, Kikuchi RKP, Lima KKM, Longo LL, Cordeiro RTS, Lira SMA, Vasconcelos SL (2009) Checklist and morphometry of benthic cnidarians from the Fernando de Noronha Archipelago, Brazil. Cah Biol Mar 50:277–290

Anderson AB, Carvalho-Filho A, Morais RA, Nunes LT, Quimbayo JP, Floeter SR (2015) Brazilian tropical fishes in their southern limit of distribution: checklist of Santa Catarina's rocky reef ichthyofauna, remarks and new records. Check List 11:1688

Anderson AB, Joyeux J-C, Floeter SR (2020a) Spatiotemporal variations in density and biomass of rocky reef fish in a biogeographic climatic transition zone: trends over 9 years, inside and outside the only nearshore no-take marine-protected area on the southern Brazilian coast. J Fish Biol 97(3):845–859

Anderson AB, Silva JP, Sorvilo R, Francini CLB, Floeter SR, Barreiros JP (2020b) Population expansion of the invasive Pomacentridae *Chromis limbata* (Valenciennes, 1833) in southern Brazilian coast: long-term monitoring, fundamental niche availability and new records. J Fish Biol 97:362–373

Andrades R, Joyeux J-C, Gasparini JL, Reis-Filho JA, Macieira RM, Giarrizzo T (2017) Fringe on the brink: intertidal reefs at risk. Science 357:261–261

Andrades R, Reis-Filho JA, Macieira RM, Giarrizzo T, Joyeux J-C (2018) Endemic fish species structuring oceanic intertidal reef assemblages. Sci Rep 8:10791

Araújo FG, Teixeira TP, Guedes APP, Azevedo MCC, Pessanha ALM (2018) Shifts in the abundance and distribution of shallow water fish fauna on the southeastern Brazilian coast: a response to climate change. Hydrobiologia 814:205–218

Araújo GS, Vilasboa A, Britto MR, Bernardi G, Von Der Heyden S, Levy A, Floeter SR (2020) Phylogeny of the comb-tooth blenny genus *Scartella* (Blenniiformes: Blenniidae) reveals several cryptic lineages and a trans-Atlantic relationship. Zool J Linnean Soc 190:54–64

Aued AW, Smith F, Quimbayo JP, Cândido DV, Longo GO, Ferreira CEL, Witman JD, Floeter SR, Segal B (2018) Large-scale patterns of benthic marine communities in the Brazilian Province. PLoS One 13(6):e0198452

Barros-Alves SDP, Alves DFR, Cobo VJ (2017) Brachyuran crab (Crustacea, Decapoda) assemblage associated with *Sargassum cymosum* in southeastern Brazil. Mar Biodivers 48:2043–2055

Barroso CX, Lotufo TMDC, Matthews-Cascon H (2016) Biogeography of Brazilian prosobranch gastropods and their Atlantic relationships. J Biogeogr 43:2477–2488

Bender MG, Pie MR, Rezende EL, Mouillot D, Floeter SR (2013) Biogeographic, historical and environmental influences on the taxonomic and functional structure of Atlantic reef fish assemblages. Glob Ecol Biogeogr 22:1173–1182

Blanchette CA, Broitman BR, Gaines SD (2006) Intertidal community structure and oceanographic patterns around Santa Cruz Island, CA, USA. Mar Biol 149:689–701

Borthagaray AI, Carranza A (2007) Mussels as ecosystem engineers: their contribution to species richness in a rocky littoral community. Acta Oecol 31:243–250

Bosa CR, Masunari S (2002) Crustáceos decápodos associados aos bancos de *Phragmatopoma caudata* (Kröyer) (Polychaeta, Sabellariidae) na Praia de Caiobá, Matinhos, Paraná. Rev Bras Zool 19:117–133

Bosman AL, Hockey PAR, Siegfried WR (1987) The influence of coastal upwelling on the functional structure of rocky intertidal communities. Oecologia 72:226–232

Brasileiro PS, Yoneshigue-Valentin Y, Bahia RDG, Reis RP, Filho GMA (2009) Algas marinhas bentônicas da região de Cabo Frio e arredores: síntese do conhecimento. Rodriguésia 60:39–66

Brayard A, Escarguel G, Bucher H (2005) Latitudinal gradient of taxonomic richness: combined outcome of temperature and geographic mid-domains effects? J Zool Syst Evol Res 43:178–188

Briggs JC (1974) Marine zoogeography. McGraw-Hill Book Company, New York, p 475p

Bumbeer J, Cattani AP, Chierigatti NB, Rocha RM (2016) Biodiversity of benthic macroinvertebrates on hard substrates in the Currais Marine Protected Area, in southern Brazil. Biota Neotrop 16(4):e20160246

Bumbeer J, Rocha RM, Bornatowski H, Robert MC, Ainsworth C (2018) Predicting impacts of lionfish (*Pterois volitans*) invasion in a coastal ecosystem of southern Brazil. Biol Invasions 20:1257–1274

Burrows M, Harvey R, Robb L (2008) Wave exposure indices from digital coastlines and the prediction of rocky shore community structure. Mar Ecol Prog Ser 353:1–12

Calado L, Silveira ID, Gangopadhyay A, Castro BD (2010) Eddy-induced upwelling off Cape São Tomé (22°S, Brazil). Cont Shelf Res 30:1181–1188

Calazans SH, Lourenço CR, Nicastro KR, Tagliaro CH, Zardi GI, Ferreira CEL, Fernandes FC, Silva EP, Hoffman EA (2021) Unlocking the history of a trans-Atlantic invader: did the human slave trade impact Brown mussel dispersal? J Biogeogr 28:2671–2681. https://doi.org/10.1111/jbi.14232

Campos EJD, Lentini CAD, Miller JL, Piola AR (1999) Interannual variability of the sea surface temperature in the South Brazil Bight. Geophys Res Lett 26:2061–2064

Campos EJD, Velhote D, Silveira ICAD (2000) Shelf break upwelling driven by Brazil Current Cyclonic Meanders. Geophys Res Lett 27:751–754

Campos TFC, Petta RA, Sichel SE, Hartmann LA, Theye T, Bezerra FHR, Srivastava NK, Mata JMLS (2009) Estado da Arte da Geologia do Arquipélago de São Pedro e São Paulo (Atlântico Norte Equatorial Brasileiro). In: Morhr LV, Castro JWA, PMS C, Alves RJK (eds) Ilhas

Oceânicas brasileiras: da pesquisa ao manejo – volume II. MMA/Secretaria de Biodiversidade e Florestas, Brasília, 542

Campos PC, Möller OO, Piola AR, Palma ED (2013) Seasonal variability and coastal upwelling near Cape Santa Marta (Brazil). J Geophys Res Oceans 118:1420–1433

Capel KCC, Segal B, Bertuol P, Lindner A (2012) Corallith beds at the edge of the tropical South Atlantic. Coral Reefs 31:75–75

Carlton JT (1996) Pattern, process, and prediction in marine invasion ecology. Biol Conserv 78:97–106

Carvalho-Filho A (1999) Peixes: costa brasileira. Editora Melro, São Paulo, pp 1–320

Carvalho-Filho A, Ferreira CE, Craig M (2009) A shallow water population of *Pronotogrammus martinicensis* (Guichenot, 1868) (Teleostei: Serranidae: Anthiinae) from south-western Atlantic, Brazil. Zootaxa 2228(1):29–42

Carvalho-Filho A, Ferreira CEL (2013) A new species of dwarf sea bass, genus *Serranus* (Serranidae: Actinopterygii), from the southwestern Atlantic Ocean. Neotrop Ichthyol 11:809–814

Carvalho-Junior L, Neves LM, Teixeira-Neves TP, Cardoso SJ (2023) Long-term changes in benthic communities following the invasion by an alien octocoral in the Southwest Atlantic, Brazil. Mar Pollut Bull 186:114386

Castro JWA (2009) Geologia Ambiental das Ilhas Oceânicas de Trindade e Fernando de Noronha, Brasil. In: Morhr LV, Castro JWA, Costa PMS, Alves RJK (eds) Ilhas Oceânicas brasileiras: da pesquisa ao manejo – volume II. MMA/Secretaria de Biodiversidade e Florestas, Brasília, p 542

Castro-Filho B, Brandini FP, Pires-Vanin AMS, Miranda LB (2006) Multidisciplinary oceanographic processes on the Western Atlantic continental shelf between 4°N and 34°S. In: Robinson AR, Brink K (eds) The sea - Vol 14A - the Global Coastal Ocean: interdisciplinary regional studies and syntheses. Harvard University Press, Cambridge, MA, pp 259–293

Christofoletti RA, Takahashi CK, Oliveira DN, Flores AA (2010) Abundance of sedentary consumers and sessile organisms along the wave exposure gradient of subtropical rocky shores of the south-west Atlantic. J Mar Biol Assoc UK 91:961–967

Christofoletti R, Almeida T, Ciotti Á (2011) Environmental and grazing influence on spatial variability of intertidal biofilm on subtropical rocky shores. Mar Ecol Prog Ser 424:15–23

Coelho JFR, Lima SMQ, Petean FF (2021) Climate change shifts abiotic niche of temperate skates towards deeper zones in the Southwestern Atlantic Ocean. bioRxiv, preprint. https://doi.org/10.1101/2021.02.17.431632

Coelho-Souza SA, López MS, Guimarães JRD, Coutinho R, Candella RN (2012) Biophysical interactions in the Cabo Frio upwelling system, southeastern Brazil. Braz J Oceanogr 60:353–365

Colares R, Machado T, Patel FM, Brunelli NS, Viana LC, Martins R, Longo PAS, Cascaes MF (2021) Rede ecológica dos gastrópodes associados ao banco de algas da Praia da Ribanceira, Santa Catarina, Brasil. Oecol Aust 25(1):47:58

Connolly SR, Roughgarden J (1998) A latitudinal gradient in Northeast Pacific intertidal community structure: evidence for an oceanographically based synthesis of marine community theory. Am Nat 151:311

Cord I, Nunes LT, Barroso CX, Freire AS, Gadig OBF, Gomes PB, Gurgel CFD, Lindner A, Mantelatto FL, Targino AKG, Floeter SR (2022) Brazilian marine biogeography: a multi-taxa approach for outlining sectorization. Mar Biol 169:61. https://doi.org/10.1007/s00227-022-04045-8

Cordeiro CAMM, Pardal A, Giménez L, Ciotti ÁM, Jenkins SR, Burrows MT, Willians G, Christofoletti RA (2024) Environmental factors have stronger effects than biotic processes in patterns of intertidal populations along the southeast coast of Brazil. Mar Environ Res 200:106646

Coutinho R (1995) Avaliação crítica das causas da zonação dos organismos bentônicos em costões rochosos. Oecologia Brasiliensis 01:259–271

Coutinho R (2004) Programa Nacional da Biodiversidade – PRONABIO - Projeto de Conservação e Utilização Sustentável da Diversidade Biológica Brasileira – PROBIO Sub-Projeto: Avaliação e

Ações Prioritárias para a Zona Costeira e Marinha. Grupo de Ecossistemas: Costões Rochosos. Guia para o licenciamento ambiental, Atividades de Sísmica Marítima na Costa Brasileira, Atividades de Perfuração de Óleo e Gás

Creed JC (2006) Two invasive alien azooxanthellate corals, *Tubastraea coccinea* and *Tubastraea tagusensis*, dominate the native zooxanthellate *Mussismilia hispida* in Brazil. Coral Reefs 25:350–350

Cruz-Motta JJ, Miloslavich P, Guerra-Castro E, Hernández-Agreda A, Herrera C, Barros F, Navarrete SA, Sepúlveda RD, Glasby TM, Bigatti G, Cardenas-Calle M, Carneiro PBM, Carranza A, Flores AAV, Gil-Kodaka P, Gobin J, Gutiérrez JL, Klein E, Krull M, Lazarus JF, Londoño-Cruz E, Lotufo T, Macaya EC, Mora C, Mora E, Palomo G, Parragué M, Pellizzari F, Retamales R, Rocha RM, Romero L (2020) Latitudinal patterns of species diversity on South American rocky shores: local processes lead to contrasting trends in regional and local species diversity. J Biogeogr 47:1966–1979

Endo CAK, Gherardi DFM, Pezzi LP, Lima LN (2019) Low connectivity compromises the conservation of reef fishes by marine protected areas in the tropical South Atlantic. Sci Rep 9:8634

Eston VR, Migotto AE, Oliveira-Filho EC, Rodrigues FJC (1986) Vertical distribution of benthic marine organisms on rocky coasts of the Fernando de Noronha Archipelago (Brazil). Bol Inst Oceanogr 34:37–53

Faveri C, Scherner F, Farias J, Oliveira EC, Horta PA (2020) Temporal changes in the seaweed flora in Southern Brazil and its potential causes. Pan-Am J Aquat 5(2):350–357

Ferreira M, Rosso S (2009) Effects of human trampling on a rocky shore fauna on the Sao Paulo coast, southeastern Brazil. Braz J Biol 69:993–999

Ferreira CEL, Floeter SR, Gasparini JL, Ferreira BP, Joyeux JC (2004) Trophic structure patterns of Brazilian reef fishes: a latitudinal comparison. J Biogeogr 31:1093–1106

Ferreira CEL, Gonçalves JEA, Coutinho R (2006) Ship hulls and oil platforms as potential vectors to marine species introduction. J Coast Res 39:1341–1346

Ferreira CEL, Luiz OJ, Floeter SR, Lucena MB, Barbosa MC, Rocha CR, Rocha LA (2015) First record of invasive lionfish (i) for the Brazilian coast. PLoS One 10(4):e0123002

Floeter SR, Soares-Gomes A (1999) Biogeographic and species richness patterns of Gastropoda on the southwestern Atlantic. Rev Bras Biol 59:567–575

Floeter SR, Guimarães RZP, Rocha LA, Ferreira CEL, Rangel CA, Gasparini JL (2001) Geographic variation in reef-fish assemblages along the Brazilian coast. Glob Ecol Biogeogr 10:423–431

Floeter SR, Ferreira CEL, Dominici-Arosemena A, Zalmon IR (2004) Latitudinal gradients in Atlantic reef fish communities: trophic structure and spatial use patterns. J Fish Biol 64:1680–1699

Floeter SR, Behrens MD, Ferreira CEL, Paddack MJ, Horn MH (2005) Geographical gradients of marine herbivorous fishes: patterns and processes. Mar Biol 147:1435–1447

Floeter SR, Rocha LA, Robertson DR, Joyeux JC, Smith-Vaniz WF, Wirtz P, Edwards AJ, Barreiros JP, Ferreira CEL, Gasparini JL, Brito A, Falcón JM, Bowen BW, Bernardi G (2008) Atlantic reef fish biogeography and evolution. J Biogeogr 35(1):22–47

Flores AAV, Negreiros-Fransozo ML (1999) On the population biology of the mottled shore crab *Pachygrapsus transversus* (Gibbes, 1850) (Brachyura, Grapsidae) in a subtropical area. Bull Mar Sci 65(1):59–73

Flores AAV, Christofoletti RA, Peres ALF, Ciotti AM, Navarrete SA (2015) Interactive effects of grazing and environmental stress on macroalgal biomass in subtropical rocky shores: modulation of bottom-up inputs by wave action. J Exp Mar Biol Ecol 463:39–48

Franco BC, Defeo O, Piola AR, Barreiro M, Yang H, Ortega L, Gianelli I, Castello JP, Vera C, Buratti C, Pájaro M, Pezzi LP, Möller OO (2020) Climate change impacts on the atmospheric circulation, ocean, and fisheries in the southwest South Atlantic Ocean: a review. Clim Chang 162:2359–2377

Freire AS, Pinheiro MAA, Karam-Silva H, Teschima MM (2011) Biology of *Grapsus grapsus* (Linnaeus, 1758) (Brachyura, Grapsidae) in the Saint Peter and Saint Paul Archipelago, Equatorial Atlantic Ocean. Helgol Mar Res 65:263–273

Freire AS, Teschima MM, Brandão MC, Iwasa-Arai T, Sobral FC, Sasaki DK, Agostinis AO, Pie MR (2021) Does the transport of larvae throughout the south Atlantic support the genetic and morphometric diversity of the Sally Lightfoot Crabs *Grapsus grapsus* (Linnaeus, 1758) and *Grapsus adscensionis* (Osbeck, 1765) (Decapoda: Grapsidae) among the oceanic islands? J Mar Syst 223:103614

Fukami H, Budd AF, Paulay G, Solé-Cava A, Chen CA, Iwao K, Knowlton N (2004) Conventional taxonomy obscures deep divergence between Pacific and Atlantic corals. Nature 427:832–835

Gomes RS, Costa PMS, Monteiro JC, Coelho ACS, Salgado NC (2006) Moluscos das Ilhas Oceânicas Brasileiras. In: Alves RJV, Castro JWA (eds) Ilhas Oceânicas Brasileiras. Da Pesquisa ao Manejo. Ministério do Meio Ambiente, Secretária de Biodiversidade e Florestas, Brasília, pp 179–198

Gomes RS, Lima FD, Barbosa JC, Spotorno-Oliveira P, Costa PMS (2017) Moluscos da Ilha Trindade e Martin Vaz. In: Sissini MN (ed) Protrindade: programa de pesquisas científicas na Ilha da Trindade. 10 anos de pesquisas. Secretaria da Comissão Interministerial para os Recursos do Mar (SECIRM), Brasília, 200p

Guilhem I, Masi B, Creed J (2020) Impact of invasive *Tubastraea* spp. (Cnidaria: Anthozoa) on the growth of the space dominating tropical rocky-shore zoantharian *Palythoa caribaeorum* (Duchassaing and Michelotti, 1860). Aquat Invasions 15:98–113

Guimaraens MAD, Coutinho R (1996) Spatial and temporal variation of benthic marine algae at the Cabo Frio upwelling region, Rio de Janeiro, Brazil. Aquat Bot 52:283–299

Hachich NF, Bonsall MB, Arraut EM, Barneche DR, Lewinsohn TM, Floeter SR (2015) Island biogeography: patterns of marine shallow-water organisms in the Atlantic Ocean. J Biogeogr 42:1871–1882

Horta PA, Oliveira-Filho ECD (2000) Macroalgas do infralitoral do sul e sudeste do Brasil: taxonomia e biogeografia

Horta PA, Amancio E, Coimbra CS, Oliveira EC (2001) Considerations on the distribution and origin of the marine macroalgal Brazilian flora. Hoehnea 28(3):243–265

Inagaki KY, Pennino MG, Floeter SR, Hay ME, Longo GO (2020) Trophic interactions will expand geographically but be less intense as oceans warm. Glob Change Biol 26(12):6805–6812

Jacobi C (1987) Spatial and temporal distribution of Amphipoda associated with mussel beds from the Bay of Santos (Brazil). Mar Ecol Prog Ser 35:51–58

Jacobucci GB, Vieira EA, Leite FPP (2018) Influence of a narrow depth gradient on the spatial structure of *Sargassum* peracarid assemblages in Southeastern Brazil. Mar Biodivers 49:1001–1011

Klôh ADS, Farrapeira CMR, Rigo APR, Rocha RM (2013) Intertidal native and introduced barnacles in Brazil: distribution and abundance. Marine Biodivers Rec 6:e102

Knoppers B, Ekau W, Figueiredo AG (1999) The coast and shelf of east and northeast Brazil and material transport. Geo-Mar Lett 19:171–178

Kocsis ÁT, Reddin CJ, Kiessling W (2018) The stability of coastal benthic biogeography over the last 10 million years. Glob Ecol Biogeogr 27:1106–1120

Laborel J (1970) Les peuplements de Madréporaires des côtes tropicales du Brésil. PhD thesis, Abidjan Université, Marseille, France

Laborel-Deguen F, Castro CB, Nunes FD, Pires DO (2019) Recifes brasileiros: o legado de Laborel. Museu Nacional, Rio de Janeiro

Lages BG, Fleury BG, Ferreira CE, Pereira RC (2006) Chemical defense of an exotic coral as invasion strategy. J Exp Mar Biol Ecol 328:127–135

Lathlean JA, Mcwilliam RA, Ayre DJ, Minchinton TE (2015) Biogeographical patterns of rocky shore community structure in south-east Australia: effects of oceanographic conditions and heat stress. J Biogeogr 42:1538–1552

Leão ZM, Kikuchi RK, Testa V (2003) Corals and coral reefs of Brazil. In: Cortes J (ed) Latin American Coral Reefs. Elsevier Science

Leão ZMAN, Kikuchi RKP, Ferreira BP, Neves EG, Sovierzoski HH, Oliveira MDM, Maida M, Correia MD, Johnsson R (2016) Brazilian coral reefs in a period of global change: a synthesis. Brazil J Oceanogr 64:97–116

Lessios HA, Kessing BD, Robertson DR, Paulay G (1999) Phylogeography of the Pantropical Sea Urchin Eucidaris in relation to land barriers and ocean currents. Evolution 53:806

Longo PADS, Mansur KFR, Leite FPP, Passos FD (2019) The highly diverse gastropod assemblages associated with *Sargassum* spp. (Phaeophyceae: Fucales) habitats. J Mar Biol Assoc UK 99:1295–1307

López M, Lavrado H, Coutinho R (2014) Structure of intertidal sessile communities before and after the invasion of *Isognomon bicolor* (C.B. Adams, 1845) (Bivalvia, Isognomonidae) in southeastern Brazil. Aquat Invasions 9:457–465

Lotufo TMC (2002) Ascidiacea (Chordata, Tunicata) do litoral tropical brasileiro. PhD thesis, Universidade de São Paulo/SP, Brazil

Luiz OJL, Ferreira CE, Rocha LA (2009) *Halichoeres sazimai*, a new species of wrasse (Perciformes: Labridae) from the Western South Atlantic. Zootaxa, 2092:37–46

Luiz OJ, Santos WCR, Marceniuk AP, Rocha LA, Floeter SR, Buck CE, Klautau AGCM, Ferreira CEL (2021) Multiple lionfish (*Pterois* spp.) new occurrences along the Brazilian coast confirm the invasion pathway into the Southwestern Atlantic. Biol Invasions 23:3013–3019

Luza AL, Aued AW, Barneche DR, Dias MS, Ferreira CEL, Floeter SR, Francini-Filho RB, Longo GO, Quimbayo JP, Bender MG (2023) Functional diversity patterns of reef fish, corals and algae in the Brazilian biogeographical province. J Biogeogr 50(6):1163–1176

MacArthur RH, Wilson EO (1967) The theory of Island biogeography. Princeton University Press, Princeton, p 215

Machado LF, Damasceno JS, Bertoncini ÁA, Tosta VC, Farro APC, Hostim-Silva M, Oliveira C (2017) Population genetic structure and demographic history of the spadefish, *Chaetodipterus faber* (Ephippidae) from Southwestern Atlantic. J Exp Mar Biol Ecol 487:45–52

Magalhãens CA, Coutinho R (1995) Distribution and zonation of three species of predatory gastropods: patterns and adaptations to wave impact in the rocky shore. Publicação Especial do Instituto Oceanográfico, São Paulo 11:123–131

Magris RA, Costa MDP, Ferreira CEL, Vilar CC, Joyeux J-C, Creed JC, Copertino MS, Horta PA, Sumida PYG, Francini-Filho RB, Floeter SR (2020) A blueprint for securing Brazil's marine biodiversity and supporting the achievement of global conservation goals. Biodivers Res 27(2):198–215

Martínez S, del Río C (2002) Las provincias malacológicas miocenas y recientes del Atlántico sudoccidental. An Biol 24:121–130

Masi BP, Macedo IM, Zalmon IR (2009) Benthic community zonation in a breakwater on the north coast of the state of Rio de Janeiro, Brazil. Braz Arch Biol Technol 52:637–646

Mazzuco ACDA, Christofoletti RA, Pineda J, Starczak VR, Ciotti AM (2015) Temporal variation in intertidal community recruitment and its relationships to physical forcings, chlorophyll-a concentration and sea surface temperature. Mar Biol 162:1705–1725

Mazzuco ACDA, Christofoletti RA, Coutinho R, Ciotti ÁM (2018) The influence of atmospheric cold fronts on larval supply and settlement of intertidal invertebrates: case studies in the Cabo Frio coastal upwelling system (SE Brazil). J Sea Res 137:47–56

Mazzuco ACDA, Stelzer PS, Bernardino AF (2020) Substrate rugosity and temperature matters: patterns of benthic diversity at tropical intertidal reefs in the SW Atlantic. PeerJ 8:e8289

McCartney MA, Keller G, Lessios HA (2000) Dispersal barriers in tropical oceans and speciation in Atlantic and eastern Pacific Sea urchins of the genus *Echinometra*. Mol Ecol 9:1391–1400

Menezes NM, McFadden CS, Miranda RJ, Nunes JACC, Lolis L, Barros F, Sampaio CLS, Pinto TK (2022) New non-native ornamental octocorals threatening a Southwest Atlantic reef. J Mar Biol Assoc UK 101(6):911–917

Menge BA (1991) Relative importance of recruitment and other causes of variation in rocky intertidal community structure. J Exp Mar Biol Ecol 146:69–100

Menge BA (2002) Environmental stress, bottom-up effects, and community dynamics: integrating molecular-physiological and ecological approaches. Integr Comp Biol 42:892–908

Menge BA, Lubchenco J, Bracken MES, Chan F, Foley MM, Freidenburg TL, Gaines SD, Hudson G, Krenz G, Leslie H, Menge DNL, Russell R, Webster MS (2003) Coastal oceanography sets the pace of rocky intertidal community dynamics. PNAS 100(21):12229–12234

Messano LVR, Gonçalves JEA, Messano HF, Campos SHC, Coutinho R (2019) First report of the Asian green mussel *Perna viridis* (Linnaeus, 1758) in Rio de Janeiro, Brazil: a new record for the southern Atlantic Ocean. Bioinvasions Rec 8(3):653–660

Micheletti-Flores C, Negreiros-Fransozo ML (1999) Porcellanid crabs (Crustacea, Decapoda) inhabiting sand reefs built by *Phragmatopoma lapidosa* (Polychaeta Sabellariidae) at Paranapuã beach, São Vicente, SP, Brazil. Rev Bras Biol 59:63–73

Miyashita LK, Melo-Junior M, Lopes RM (2009) Estuarine and oceanic influences on copepod abundance and production of a subtropical coastal area. J Plankton Res 31(8):815–826

Moraes FC, Muricy GRS (2017) Esponjas da Ilha da Trindade. In: Sissini MN (ed) Protrindade: programa de pesquisas científicas na Ilha da Trindade. 10 anos de pesquisas. Secretaria da Comissão Interministerial para os Recursos do Mar (SECIRM), Brasília, 200p

Moraes FC, Batista D, Reis F, Monteiro L, Oliveira M, Muricy G (2009) As esponjas: passado, presente e futuro. In: Hazin FHS (ed) O Arquipélago de São Pedro e São Paulo: 10 anos de Estação Científica. Secretaria da Comissão Interministerial para os Recursos do Mar (SECIRM), Brasília, p 348

Morais RA, Ferreira CEL, Floeter SR (2017) Spatial patterns of fish standing biomass across Brazilian reefs. J Fish Biol 91:1642–1667

Moura RLD, Gasparini JL, Sazima I (1999) New records and range extensions of reef fishes in the Western South Atlantic, with comments on reef fish distribution along the Brazilian coast. Rev Bras Zool 16:513–530

Muricy G, Lopes DA, Hadju E, Carvalho MS, Moraes FC, Klautau M, Menegola C, Pinheiro U (2011) Catalogue of Brazilian Porifera. Museu Nacional, Rio de Janeiro, p 300

Murray SN, Dethier MN, Ambrose RF (2006) Monitoring rocky shores. University of California Press, p 240

Neves EG, Andrade SCS, Silveira FLD, Solferini VN (2007) Genetic variation and population structuring in two brooding coral species (*Siderastrea stellata* and *Siderastrea radians*) from Brazil. Genetica 132:243–254

Nunes FLD, Norris RD, Knowlton N (2011) Long distance dispersal and connectivity in Amphi-Atlantic Corals at regional and basin scales. PLoS One 6:e22298

Nunes JACC, Sampaio CLS, Barros F (2013) How wave exposure, group size and habitat complexity influence foraging and population densities in fishes of the genus *Halichoeres* (Perciformes: Labridae) on tropical rocky shores. Mar Biol 160:2383–2394

Nunes FLD, Wormhoudt AV, Faroni-Perez L, Fournier J (2016) Phylogeography of the reef-building polychaetes of the genus *Phragmatopoma* in the western Atlantic Region. J Biogeogr 44:1612–1625

O'Connor NE, Donohue I, Crowe TP, Emmerson MC (2011) Importance of consumers on exposed and sheltered rocky shores. Mar Ecol Prog Ser 443:65–75

Oliveira OS, Martins-Silva MJ, Barros JCN, Swoboda I, Francisco JA, Schneider M, Batista JCL, Aquino PPU, Vianna G, Coutinho MF (2009) Malacofauna. In: Hazin FHS (ed) O Arquipélago de São Pedro e São Paulo: 10 anos de Estação Científica. Secretaria da Comissão Interministerial para os Recursos do Mar (SECIRM), Brasília, p 348

Oricchio FT, Marques AC, Hajdu E, Pitombo FB, Azevedo F, Passos FD, Vieira LM, Stampar SN, Rocha SM, Dias GM (2019) Exotic species dominate marinas between the two most populated regions in the southwestern Atlantic Ocean. Mar Pollut Bull 146:884–892

Palácio FJ (1982) Revisión zoogeográfica marina del sur del Brasil. Bol Inst Oceanogr 31:69–92

Palóczy A, Silveira ID, Castro B, Calado L (2014) Coastal upwelling off Cape São Tomé (22°S, Brazil): the supporting role of deep ocean processes. Cont Shelf Res 89:38–50

Pardal A, Cordeiro CAMM, Ciotti AM, Jenkins SR, Gímenez L, Burrows MT, Christofoletti RA (2021) Influence of environmental variables over multiple spatial scales on the population structure of a key marine invertebrate. Mar Environ Res 179:105410

Pardal A, Martinez AS, Ciotti AM, Christofoletti RA, Cordeiro CAMM (2023) Macroecology of rocky intertidal benthic communities along the southwestern Atlantic: patterns of spatial variation and associations with natural and anthropogenic variables. Mar Environ Res 190:106099

Paula JCD, Coração AC, Lopes-Filho EA, Silva RP, Santos LND, Carvalho WFD (2020) Diversity and turnover in a rocky shore intertidal community of an upwelling region (Arraial do Cabo, Brazil). Anais Acad Bras Cienc 92(2):e20181096

Paz JRL, Oliveira MJ, Silva PP (2012) Levantamento da Malacofauna bentônica de costões rochosos do mesolitoral da Praia de Plataforma, Bahia. Candombá Revista Virtual 8(1):72–80. (Portuguese only)

Peluso L, Tascheri V, Nunes FLD, Castro CB, Pires DO, Zilberberg C (2018) Contemporary and historical oceanographic processes explain genetic connectivity in a Southwestern Atlantic coral. Sci Rep 8(1):2684

Picciani N, de Lossio e Seiblitz IG, de Paiva PC, e Castro CB, Zilberberg C (2016) Geographic patterns of *Symbiodinium* diversity associated with the coral *Mussismilia hispida* (Cnidaria, Scleractinia) correlate with major reef regions in the Southwestern Atlantic Ocean. Mar Biol 163:236

Pinheiro HT, Bernardi G, Simon T, Joyeux JC, Macieira RM, Gasparini JL, Rocha C, Rocha LA (2017) Island biogeography of marine organisms. Nature 549(7670):82–85

Pinheiro HT, Rocha LA, Macieira RM, Carvalho-Filho A, Anderson AB, Bender MG, Dario FD, Ferreira CEL, Figueiredo-Filho J, Francini-Filho R, Gasparini JL, Joyeux J-C, Luiz OJ, Mincarone MM, Moura RL, José De Anchieta CC, Nunes QJP, Rosa RS, Sampaio CLS, Sazima I, Simon T, Vila-Nova DA, Floeter SR (2018) South-western Atlantic reef fishes: zoogeographical patterns and ecological drivers reveal a secondary biodiversity Centre in the Atlantic Ocean. Divers Distrib 24:951–965

Quimbayo JP, Dias MS, Kulbicki M, Mendes TC, Lamb RW, Johnson AF, Aburto-Oropeza O, Alvarado JJ, Bocos AA, Ferreira CEL, Garcia E, Luiz OJ, Mascareñas-Osorio I, Pinheiro HT, Rodriguez-Zaragoza F, Salas E, Zapata FA, Floeter SR (2018) Determinants of reef fish assemblages in tropical Oceanic islands. Ecography 42:77–87

Reid DG (2009) The genus *Echinolittorina* Habe, 1956 (Gastropoda: Littorinidae) in the western Atlantic Ocean. Zootaxa 2184:1–103

Rios EC (1994) Seashells of Brazil. Rio Grande, Museu Oceanográfico Prof. E.C. Rios da Fundação Universidade de Rio Grande, 2nd ed., 368p

Rocha LA (2003) Patterns of distribution and processes of speciation in Brazilian reef fishes. J Biogeogr 30:1161–1171

Rocha RM, Moreno TR, Metri R (2005) Ascídias (Tunicata, Ascidiacea) da Reserva Biológica Marinha do Arvoredo, Santa Catarina, Brasil. Rev Bras Zool 22(2):461–476

Rocha LA, Rocha CR, Roberston RD, Bowen BW (2008) Comparative phylogeography of Atlantic reef fishes indicates both origin and accumulation of diversity in the Caribbean. BMC Evol Biol 8:157

Rocha RM, Dias GM, Lotufo TMC (2011) Checklist das ascídias (Tunicata, Ascidiacea) do Estado de São Paulo, Brasil. Biota Neotrop 11(1):749–759

Santana EFC, Mies M, Longo GO, Menezes R, Aued AW, Luza AL, Bender MG, Segal B, Floeter SR, Francini-Filho RB (2023) Turbidity shapes shallow Southwestern Atlantic benthic reef communities. Mar Environ Res 183:105807

Santos EC (2020) Variação espaço-temporal da estrutura das comunidades macrobentônicas em costões rochosos tropicais (Bahia-Brasil). PhD Thesis, Universidade Federal de Pernambuco, Programa de Pós-Graduação em Oceanografia

Santos S, Hrbek T, Farias IP, Schneider H, Sampaio I (2006) Population genetic structuring of the king weakfish, *Macrodon ancylodon* (Sciaenidae), in Atlantic coastal waters of South America: deep genetic divergence without morphological change. Mol Ecol 15:4361–4373

Santos MEA, Kitahara MV, Lindner A, Reimer JD (2016) Overview of the order Zoantharia (Cnidaria: Anthozoa) in Brazil. Mar Biodivers 46:547–559

Santos MEA, Faria-Junior E, Aued AW, Peluso L, Kitahara MV, Pires DO, Zilberberg C (2020) Benthic Cnidaria community in the oceanic archipelago of Trindade and Martin Vaz, Southwestern Atlantic Ocean. Reg Stud Mar Sci 33:100895

Schiel DR (2004) The structure and replenishment of rocky shore intertidal communities and biogeographic comparisons. J Exp Mar Biol Ecol 300:309–342

Schubart CD, Cuesta JA, Felder DL (2005) Phylogeography of *Pachygrapsus transversus* (Gibbes, 1850): the effect of the American continent and the Atlantic Ocean as gene flow barriers and recognition of *Pachygrapsus socius* Stimpson 1871 as a valid species. Nauplius 13(2):99–113

Silva AGD, Paula AFD, Fleury BG, Creed JC (2014) Eleven years of range expansion of two invasive corals (*Tubastraea coccinea* and *Tubastraea tagusensis*) through the southwest Atlantic (Brazil). Est Coast Shelf Sci 141:9–16

Simone LRL (2008) A New Species of *Fissurella* from São Pedro and São Paulo Archipelago, Brazil (Vetigastropoda, Fissurellidae). Veliger 50(4):292–304

Sissini MN, Oliveira MC, Horta PA, Pellizzari F (2017) Macroalgas da Ilha da Trindade. In: Sissini MN (ed) Protrindade: programa de pesquisas científicas na Ilha da Trindade. 10 anos de pesquisas. Secretaria da Comissão Interministerial para os Recursos do Mar (SECIRM), Brasília, p 200

Sissini MN, Koerich G, de Barros-Barreto MB et al (2021) Diversity, distribution, and environmental drivers of coralline red algae: the major reef builders in the Southwestern Atlantic. Coral Reefs 41:711–725

Soares MO, Lemos VB, Kikuchi RKP (2011) Aspectos biogeomorfológicos do Atol das Rocas, Atlântico Sul Equatorial. R Bras Geoc 41(1):85–94

Soares MO, Feitosa CV, Garcia TM, Cottens KF, Vinicius B, Paiva SV et al (2022) Lionfish on the loose: *Pterois* invade shallow habitats in the tropical southwestern Atlantic. Front Mar Sci 9:956848

Souza RCCL, Lima TA, Silva EP (2010) Holocene molluscs from Rio de Janeiro state coast, Brazil. Check List 6:301

Spalding MD, Fox HE, Allen GR, Davidson N, Ferdaña ZA, Finlayson M, Halpern BS, Jorge MA, Lombada A, Lourie SA, Martin KD, McManus E, Molnar J, Recchia CA, Robertson J (2007) Marine ecoregions of the world: a bioregionalization of coastal and shelf areas. Biosci 57(7):573–583

Steigleder KM, Copertino MS, Lanari M, Camargo M, Fujii MT (2019) Latitudinal gradient in intertidal seaweed composition off the coast of southern Brazil and Uruguay. Aquat Bot 156:47–56

Széchy MTM, Veloso VG, Paula EJ (2001) Brachyura (Decapoda, Crustacea) of phytobenthic communities of the sublittoral region of rocky shores of Rio de Janeiro and São Paulo, Brazi. Trop Ecol 42(2):231–242

Teixeira L, Creed J (2020) A decade on: an updated assessment of the status of marine non-indigenous species in Brazil. Aquat Invasions 15:30–43

Teschima MM, Ströher PR, Firkowski CR, Pie MR, Freire AS (2016) Large-scale connectivity of *Grapsus grapsus* (Decapoda) in the Southwestern Atlantic oceanic islands: integrating genetic and morphometric data. Mar Ecol 37:1360–1372

Trovant B, Ruzzante DE, Basso NG, Orensanz JL (2013) Distinctness, phylogenetic relations and biogeography of intertidal mussels (*Brachidontes*, Mytilidae) from the south-western Atlantic. J Mar Biol Assoc UK 93:1843–1855

Trovant B, Basso NG, Orensanz JM, Lessa EP, Dincao F, Ruzzante DE (2016) Scorched mussels (*Brachidontes* spp., Bivalvia: Mytilidae) from the tropical and warm-temperate southwestern Atlantic: the role of the Amazon River in their speciation. Ecol Evol 6:1778–1798

Twist B, Kluibenschedl A, Pritchard D, Desmond M, D'Archino R, Nelson W, Hepburn C (2020) Biomass and species richness relationships in macroalgal communities that span intertidal and subtidal zones. Mar Ecol Prog Ser 654:67–78

Valentin JL (2001) The Cabo Frio upwelling system, Brazil. In: Seeliger U, Kjerfve B (eds) Coastal marine ecosystems of Latin America. Ecological studies, vol 144. Springer, Berlin/Heidelberg

van Soest R, Hajdu E (1997) Marine area relationships from twenty Sponge Phylogenies. A comparison of methods and coding strategies. Cladistics 13:1–20

Villac MC, Tenenbaum DR (2010) The phytoplankton of Guanabara Bay, Brazil. I. historical account of its biodiversity. Biota Neotrop 10(2):271–293

Villaça R, Pedrini AG, Pereira SMB, Figueiredo MAO (2006) Flora marinha bentônica das ilhas oceânicas brasileiras. In: Alves RJV, Castro JWA (eds) Ilhas oceânicas brasileiras da pesquisa ao manejo. Ministério do Meio Ambiente – SBF, Brasília, pp 105–146

Wares JP (2019) Small, flat, and gray: cryptic diversity in chthamalid barnacles in the global context of marine coastal biogeography (Cirripedia: Balanomorpha: Chthamalidae). J Crustac Biol 40:1–16

Young PS (1993) The Verrucomorpha and Chthamaloidea from the Brazilian coast (Crustacea: Cirripedia). Rev Bras Biol 53(2):255–267

Chapter 3
Benthic-Pelagic Coupling

Lohengrin D. A. Fernandes and Bernardo A. P. da Gama

Abstract Marine organisms living on rocky shores are not isolated in the benthic realm. They are instead intrinsically connected with the surrounding pelagic realm, and many distinct links between them could exemplify this interdependency. One of the most fundamental aspects is reproduction, as they more often release numerous pelagic larvae that must return to a benthic lifestyle at the end of their larval development. All these cyclical links that exemplify the interdependency between benthic and pelagic environments can be briefly described in terms of the exchange of energy and matter, the so-called benthic-pelagic coupling. Since the early 1950s, BP coupling has usually been addressed in terms of connections between coastal rocky shores and the shelf, particularly in subtropical southeastern and southern Brazil. Therefore, most studies focusing on BPC in rocky shores address ecosystems dominated by subtropical species, mainly crustaceans, mollusks, polychaetes, and echinoderms. In this chapter, we provide a theoretical background on the BPC from a new perspective, focusing on the anthropogenic interferences registered in the Southwestern Atlantic Ocean over the last decades. We also review classical concepts of larval dynamics, development, recruitment, and physiological constraints.

Introduction

Most marine benthic organisms living on rocky shores and other benthic ecosystems benefit from the pelagic environment in different ways, like releasing their gametes, offspring, and excreta in the water. There are many advantages in doing so, such as extending the ability to spread either desired or undesired matter to larger areas, reducing competition between adults and non-adults, and promoting the

L. D. A. Fernandes (✉)
Departamento de Biotecnologia Marinha, Instituto de Estudos do Mar Almirante Paulo Moreira, Arraial do Cabo, RJ, Brazil
e-mail: lohengrin@marinha.mil.br

B. A. P. da Gama
Departamento de Biologia Marinha, Universidade Federal Fluminense, Niterói, RJ, Brazil

R. Coutinho, R. A. Christofoletti (eds.), *Brazilian Rocky Shores*, Brazilian Marine Biodiversity, https://doi.org/10.1007/978-3-031-67206-4_3

genetic flow between distant populations. The classical example starts with a benthic adult releasing many pelagic larvae that in turn must return to a benthic lifestyle (benthic-pelagic pathway). Contrarily, pelagic adults like fishes and squids may also benefit from a benthic environment by concentrating their gametes or protecting their offspring in a benthic nest or burrow. By delivering their DNA-bearing forms in a less dispersive environment, organisms that use this pelagic-benthic pathway reduce the loss and consequently increase the individual chance of success. After this benthic phase, the cycle ends with a return to the pelagic world where adults can survive and spread. In common to both pathways, marine species must get through constraints in their life cycles that usually affect population growth. Anyway, both pathways are essentially the same cyclical link between benthic and pelagic environments involving the exchange of energy and matter, collectively named benthic-pelagic coupling or BPC (Griffiths et al. 2017).

The benthic-pelagic coupling is more often described worldwide in terms of connections between (1) mangrove and estuary or (2) rocky shore and shelf. Particularly in Brazil (southwestern Atlantic), larger mangroves are nowadays located in the tropical northern and northeastern regions, while rocky shores are frequent in the subtropical southeastern and southern regions (Villwock et al. 2005; Coutinho et al. 2016). Therefore, most studies focusing on BPC in rocky shores address subtropical ecosystems and species, mainly larvae of barnacles and mussels.

The majority of marine benthic organisms are biphasic strategists that thrive in both realms, benthic and pelagic (Pechenik 1999). The evolutionary mechanisms that selected this complex life cycle were likely related to the dispersion of larvae and propagules, even though some evidence suggests other selective forces as equally possible (Swearer et al. 2002; Strathmann and Grunbaum 2006). For example, larval development in the plankton could be either occasional or a strategy to minimize predation risk, increase nourishment, minimize competition with adult conspecifics, disperse to new environments, or even break parasite cycles. All these selective forces could have simultaneously driven the evolution of many distinct lifecycles as currently seen in the multiple meroplanktonic living forms (i.e., living as plankton during part of the life history).

In the past, there was a general thought that metamorphosis is one of the most dramatic events in animal development, splitting the life cycle into at least two phases, namely, "larval" and "post-larval" (Stanley et al. 2015; Mikhlina et al. 2020; Pepe-Victoriano et al. 2021). Others instead prefer to adopt a more ecological approach and use the settlement to split these two phases, thereby adopting terms like "pre-" and "post-settlement," "pre-" and "post-juvenile," or "pre-competent" and "competent" to settle (Scheltema 1986; Amieva and Reed 1987; Menge 1991; Wilson and Harrison 1998; Gallego et al. 2012). Besides metamorphosis and settlement, other terms like "hatching" and "sexual maturity" are also milestones in development that can split embryonic from larval, and juvenile from adult phases, for example. Thus, general development can be broadly described in four phases: embryo, larva, juvenile, and adult (Fig. 3.1). A general and useful definition of terms, describing some basic concepts in larval morphology and ecology, are presented by different authors (McEdward and Janies 1993; Anger 2002; Fraschetti et al. 2003; López and Coutinho 2008).

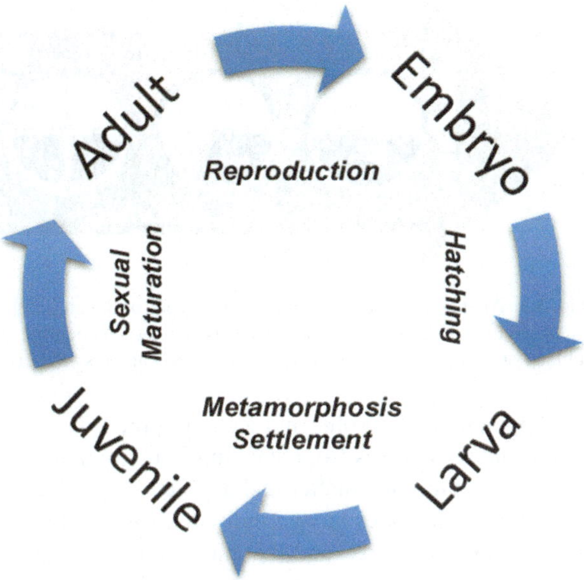

Fig. 3.1 A generalized sequence of phases in the development of a hypothetic benthic-pelagic organism and the most important events that separate them

Larval Development

Anamorphic and Metamorphic Developments

It was not always an easy task to distinguish between the developmental phases of one species. What can be seen as an obvious task today has led researchers in the past centuries to some misidentification. Many crustacean larvae, for example, were first described as adult forms belonging to new genera, like *Nauplius* O.F. Müller, 1785, *Cypris* O.F. Müller, 1776, and *Megalopa* Leach, 1813. It took some time until Thompson's work (1828) proved that most benthic crustacean species undergo a larval planktonic phase and a metamorphosis to acquire their final morphology. Until now, there is no single description that easily summarizes all kinds of development among marine organisms, but from an ecological perspective, it is useful to start splitting the lifecycle into a non-reproductive (or immature) phase followed by a reproductive (mature) one (Fig. 3.2). From a very basic perspective, there are no morphological differences between these two phases, so their coexistence in two distinct realms would benefit a species by reducing competition for food and space. For instance, some benthic living bacteria can move to a pelagic lifestyle by switching on/off specific genes and performing an adaptive response. The return to a benthic substrate is usually associated with a biofilm formation that ends this biphasic lifecycle and creates a microbial benthic-pelagic coupling (Sousa et al. 2011). This basic development can evolve to a more complex model in protozoans and metazoans, which includes an additional stage in the lifecycle like a cyst or an egg. Depending on the morphology (bauplan) of what hatches, there are two sorts of

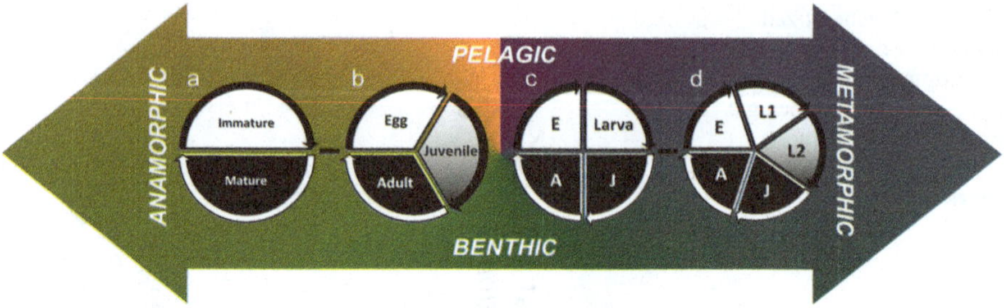

Fig. 3.2 Anamorphic versus metamorphic types of development, from 2 and 3 phases anamorphic (**a** and **b**) to 4 and 5 phases metamorphic (**c** and **d**). The benthic and pelagic phases are marked with a black or white background, respectively, and can be switched. *E* egg, *L* larva, *J* juvenile, *A* adult

development: anamorphic and metamorphic (Fig. 3.2). In the first case, there are no larvae and what arises from hatching is an immature juvenile, morphologically similar to the adult, but smaller and unable to reproduce. In contrast, the metamorphic development includes larvae: individuals whose bauplan is distinct from that of the adult besides their size, and that must undergo throughout metamorphosis. In some species' development, there are multiple larval stages (larval stage 1 → larval stage 2 → larval stage n) before the juvenile.

Since no organism that leaves the egg is completely identical to the adult, there is no consensus on the cutting point between anamorphic and metamorphic development. In other words, the type of development is somewhat arbitrary. For example, crustacean isopods and amphipods are thought to have anamorphic development because the juvenile exhibits the adult bauplan since hatching. In contrast, metamorphic development includes many taxa, like echinoderms, mollusks, and crustacean cirripeds, whose larvae are hardly ascribed to species level by inexperienced taxonomists. This complex life cycle is more often related to the benthic-pelagic coupling since epibenthic organisms are less studied.

Larval Bauplan

Benthic organisms that live associated with rocky shores usually have 4 or 5 phases in metamorphic development that include one stage specialized in dispersion and another in settlement. It means that in each stage, the larvae must fulfill all morphological, physiological, and behavioral requirements to thrive until the next phase. The adoption of a 5-stage biphasic development, with one dispersion-specialized (e.g., nauplius and zoea) and other settlement-specialized (e.g., cypris and megalopae), implicates two distinct larval bauplans for a species and an energetic commitment with massive restructuring. Despite the energetic cost of such specific requirements, many unrelated taxons living on rocky shores adopt this type of development, suggesting that it must bring some advantage to the recruitment compared to those species with a single generalist larva (4-stage biphasic). Thus, the first stages, such as nauplii of barnacles, trochophores of mussels, and zoeae of crabs, hatch ready to survive for long periods as either feeding planktonic organisms (planktotrophic) or non-feeding (lecithotrophic) but are unable to settle and/or attach to the rocky shore. In contrast, the last stages, like (non-feeding) cypris, pediveliger, and megalopae, are not well suitable to survive for a long time in the plankton. They have instead stronger sensitive appendages—like antennae, feet, and chelipeds, and adhesive glands (the cement and byssal glands) that help them to search for and attach to an appropriate benthic substrate. Any adjustment in the larval bauplan to achieve the specific requirements to survive and thrive increases the complexity in the larval development and thus the energetic costs.

Morphological adjustments can be sometimes gradual between stages. The first stage that emerges from a barnacle's egg—nauplius I—promptly starts its journey toward the inner shelf bearing three appendages with a certain number of setae in each. As the larva grows and molts through stages II and so, the number of setae increases to improve its ability to swim and/or to feed relative to an increase in the larval size and energetic requirement. Otherwise, the appendages would not generate enough power for the larva to move forward or bring enough food for them to survive. This is a general rule for the initial stages of any larvae, like trochophore and early veliger (D-larva) of mussels (e.g., *Mytilus trossulus* Dyachuk and Odintsova 2009), and zoea of crabs in the subtidal zone. Besides the number of setae, new appendages and body parts may arise during larval development, dramatically changing the larval behavior.

Besides rocky shores, this n-stage biphasic model of development is spread over distinct marine ecosystems, like estuaries, bays, and coral reefs, where the meroplankton is composed of multiple larval shapes like the planned phyllosoma larvae of lobster and the V-shaped echinopluteus larvae of sea urchin. These larvae have specific bauplan adapted to the environment they live in, which in turn affect survival rate, settlement, and recruitment. Phyllosome larvae, for instance, live floating in the open ocean for long periods, differently from an ascidian tadpole larva well adapted to a short lifetime near the coast. Thus, the larval bauplan that leads to a higher larval survival rate is positively selected in each population, although there is also evidence on the contrary (Underwood and Fairweather 1989). The exact mechanisms by which a larval phenotype is selected in the population are still under study, but some evidence reveals a relationship between larval bauplan and recruitment. Taking into account the high influence of local biotic and abiotic factors upon recruitment (Vaughn and Allen 2010), the larval bauplan would be highly important for speciation. For instance, those larvae that benefit from reduced mortality likely reach the shoreline in higher densities and more often saturate, thus succeeding in recruitment. Phyllosomae with reduced pigmentation are less suitable for predation by a visual hunter like planktivorous fish, but in contrast, they are more susceptible to mortality from UV radiation. There exists a trade-off between UV protection and camouflage, but in general, any larvae that exhibit colorful pigmentation are less transparent and therefore more consumed (Hansson 2000). The shells of D-larvae of mussels are less opaque than those of pediveligers, which has also pigmented eyes

Table 3.1 Distribution of larval bauplan among bilateral and radial symmetry

Bilateral	*Triangular*: Trochophore (Polychaeta), Nauplius (Cirripedia), Cyphonautes (Bryozoa), Pilidium (Nemertean)
	Elliptical: D-larvae (Mollusca), Planula (Coral), Nauplius (Copepoda), Veliger, Cypris, Tornaria (Hemichordata), Bipinnaria, Pentacula (Holothuroidea), Actinotrocha (Phoronida)
	Elongate: Mysis (Crustacean), Tadpole (Ascidia), Fish-larvae
Radial	*Spherical, globular, or round*: Auricularia (Holothuroidea)
	Spiny: Echinopluteus (Echinodermata), Ophiopluteus (Echinodermata)

Adapted from Smith and Johnson (1996)

(Satuito et al. 1994). Then, small intraspecific or interspecific differences between larval pigmentation might affect the survival rate by increasing or decreasing protection against predation (Bashevkin et al. 2020). Strategies to deter predation among planktonic larvae include an almost complete lack of pigmentation (eyes are usually the only visible features underwater) or transparency, and long body projections such as spicules or spines (McClintock et al. 1996). These features also help increase friction and thus act as a sort of "parachute" reducing sinking and facilitating long-distance transport by ocean currents. Different larval bauplans are found among bilateral and radial taxa (Table 3.1) that fit into the strategies adopted to increase buoyancy, mobility, or protection. Many species from rocky shores exhibit spines or hard parts to prevent predation. The V-shaped echinopluteus of sea urchins, for example, may have 2 to 8 spinous arms. Unlike other nauplii (e.g., Copepoda, Euphausiacea, and Decapoda nauplii), barnacle nauplii have spines on the carapace margin. Similarly, crab zoeae that undergo development over the continental shelf have larger spines compared to zoeae from species restricted to mangroves and estuaries (Anger 2002). Bivalvia D-larvae (e.g., Ostreidae and Mytilidae) are born with a thin shell where the larvae can hide and protect. As a general rule, the less protected the more consumed the larvae, even though there is evidence suggesting the larval size rather than spines as a primary selective force for planktivorous feeding (Morgan 1992). The need for protection is so crucial for larval survival on rocky shores that some larvae can even anticipate adult skills, like the ability to build a sandy tube at the time of metamorphosis, as observed in some sabellarid polychaetes (Amieva and Reed 1987).

Planktotrophy and Lecitothrophy

Besides avoiding mortality by predation after hatching, any larva must grow and metamorphose to succeed in recruitment. The energy needed is therefore proportional to the increase in body mass during growth and the complexity of metamorphosis. Some species have free-swimming larvae that hatch with all the energy needed to grow and metamorphose, named lecithotrophic, while others (planktotrophic) have to feed during the pelagic phase to attain the size and maturity necessary.

Nevertheless, not all rocky shore organisms have planktonic larvae. Some hard-bottom predators, like the Brazilian gastropod *Pleuroploca aurantiaca*, have an intracapsular veliger that metamorphoses to juvenile before hatching, thus without a planktonic phase (Meirelles and Matthews-Cascon 2005). Many Brazilian gastropods lost their free-swimming veliger (Gallardo and Penchaszadeh 2001) and have instead intracapsular metamorphosis, direct development, or a free non-pelagic stage (Mileikovsky 1971), disconnected from the benthic-pelagic coupling. This type of disconnection from the BP coupling is a strategy adopted by poecilogonic species, like some spionid polychaetes and mollusk nudibranchs that alternate between both developmental modes. In this case, there is a proportion of coexisting feeding and non-feeding larvae in a larval pool strategically connected to the benthic-pelagic coupling (Vance 1973). In general, barnacle larval development begins with a lecithotrophic larva (nauplius I) that turns into planktotrophic in stage II, returning to a non-feeding final stage that allocates most of the stored energy for searching, attachment, and metamorphosis.

There are several ecological implications of the adopted developmental mode upon populational dynamics, including generation time, number of offspring, dispersion, settlement, and the adult size classes (Todd 1979; Emlet 1986; Levin et al. 1987). In general, species with planktotrophic larvae are r-strategists and have shorter generation times, numerous small eggs, and few coexisting cohorts. In contrast, lecithotrophic larvae are usually found in k-strategist species, with longer generation times, a proportionally few large eggs, plenty of yolk, and many coexisting cohorts (Thorson 1950). This general rule was first suggested for all species spreading along the latitudinal gradient of temperature despite the growing number of exceptions (Gallardo and Penchaszadeh 2001).

For an adult living on the rocky shore, how much energy can be successfully allocated to produce one viable egg or larva? To answer this question, we must take into account that the energetic cost of one larva increases proportionally to the decrease in the number of eggs produced per individual. Thus, the energy needed to produce one large lecithotrophic egg can be tenfold higher than that needed to produce one small planktotrophic egg (Todd 1979). There is a trade-off between the quality and quantity of larvae produced yearly by each species that results from developmental mode and adult nourishment. For instance, the same number of eggs can be produced in a given time by Chthamalid barnacles compared to larger Balanids, but at a lower cost (Barnes and Barnes 1965). The proportion of planktotrophic versus lecithotrophic larvae is therefore adjusted every reproductive season to reflect the energetic budget allocated for gonadal development (reproductive effort) and the time needed for embryonic and larval development. Gonadal development, including embryogenesis, and larval development are fueled by higher phytoplanktonic production and higher temperatures that occur, for example, during the spring and summer seasons in southern and southeastern Brazil. Additionally, some benthic species, like those poecilogonic spionid polychaetes that adopt a lecithotrophic mode, may have increased reproductive activity, egg production, and survivorship under high food availability (Levin and Creed 1986). Despite the adopted developmental mode, the success in the replenishment of the adult population is likely the final result of the complex set of pressures and restrictions placed upon the larval pool (Fernandes et al. 2017).

Benthic-Pelagic Dynamics

Synchronism and Larval Supply

The delivery of gametes, eggs, and larvae (GEL) to the marine environment can be either an occasional and self-dependent moment or a synchronous and populational-mediated event. In the last case, most individuals in a local population are engaged at the relatively same time—hours to a few days—to release their GEL in one to few pulses. This is called "Reproductive Synchrony" and has major impacts firstly on reproductive success and then on benthic-pelagic coupling (Ims 1990) due to its predominance among the dominant rocky shore populations: barnacles and mussels. Synchronism has evolved through generations to include larval settlement and recruitment.

Synchronism in reproduction is a common strategy among marine organisms and two hypotheses were proposed to explain advantages. The (1) "avoidance hypothesis" states that the lower the predation rate, the lesser the mortality of gametes, eggs, and larvae. Differently, the (2) "swamping hypothesis" suggests that predators get quickly satiated and stop eating when offspring is abundant (Ims 1990; Morgan and Christy 1997). In common, both hypotheses search for evolutive advantages that could explain why many rocky shore species synchronize their reproduction. For one species to be considered synchronous in the larval supply, the majority of adult mature individuals in a population must release their GELs at the relatively same time. At least four major factors are known to influence synchronism in marine ecosystems either alone or combined: (1) the seasonal change in water temperature (Skinner and Coutinho 2002; Fernandes et al. 2017), (2) the gravitational (tide) and the atmospheric (wind-driven) waves (Fernandes et al. 2002; Bueno and Flores 2010), (3) the seasonal spring bloom of phytoplankton (Junqueira et al. 1997; Ventura et al. 1997; Fernandes et al. 2017), and (4) the seasonal wet/dry rhythm (Diele and Simith 2006). During nocturnal high tides, most intertidal organisms in rocky shores, mangroves, and estuaries release their GELs in an attempt to simultaneously avoid predation (avoidance hypothesis) and promptly reach deeper areas. For instance, most larvae of *Ucides cordatus* (Decapoda) that hatch in the inner region of the Caeté estuary, Northeast Brazil, is quickly exported by offshore currents during nocturnal spring tide during the wet season (Diele and Simith 2006). Tide is also very relevant to control the larval distribution of *Anchovia clupeoides* (Clupeiformes) in the same region (Barletta and Barletta-Bergan 2009), where the water temperature is almost constant over the year and thus less significant as a cue for synchronism. The vast majority of "crab" larvae (including swimming crabs, hermit crabs, and porcelain crabs) hatch simultaneously at night during the high tide, either in Northern (Caeté estuary, Brazil) and Northeastern (Itamaracá estuarine system, Brazil) regions of Atlantic South America, where the water temperature is constant over the year, or in Southeastern (Guanabara Bay and Sepetiba Bay, Brazil) and Southern (Patos Lagoon, Brazil and Mar Chiquita Lagoon, Argentina) regions (Fernandes et al. 2002; Vuichard et al. 2013; Melo Júnior et al. 2016) that have a seasonal temperature change. Since almost all the four drivers in synchronism are seasonally variable, the phenology of most species in shallow marine ecosystems in Brazil fits into the same pattern of one to two annual peaks in larval supply. The annual peak of larval release for many soft-bottom species (e.g., swimming crabs, clams, and oysters) in shallow areas, like mangroves and estuaries, is driven by a combination of all factors: tide, temperature, food, and runoff. For those living on rocky shores, like barnacles, mussels, sea urchins, and crabs, all but runoff are relevant. Finally, for subtidal benthic organisms, like some annelids and echinoderms, temperature and food availability are the major cues (Junqueira et al. 1997; Ventura et al. 1997; Bueno and Flores 2010).

It is noteworthy that benthic marine macroalgae also produce and release gametes (haploid) or spores (diploid) in the water, where fertilization then occurs. At least in some species (e.g., brown algae such as *Fucus*), the synchronism is driven by sex pheromones released in the water by the female gametes, such as the natural hydrocarbon pheromone fucoserratene. Unfortunately, some petroleum hydrocarbons can mimic the chemotactic effect of seaweed pheromones, driving algal spermatozoids toward oil pollution instead of the female gametes, thus hindering fertilization and therefore reproduction (Derenbach and Gereck 1980).

Larval Transport, Dispersion, and Retention

Like most marine planktonic organisms, pelagic eggs from benthic organisms are analogous to passive particles suspended in the water, and therefore subject to transport by currents (McManus and Woodson 2012). Larvae, in contrast, can behave simultaneously as floating (passive) and motile (active) particles in the sea, depending on the direction of movement. The majority actively swim up and down in a vertical migration pattern that follows a tidal, daily, or ontogenetic rhythm (Tapia and Pineda 2007), but very few can actively prevent horizontal displacement and, therefore, are passively subject to transport by currents. In combination with larval behavior, three major physical forces are known to promote most of the larval dynamics: wave heights, tidal currents, and wind-driven currents (Pineda et al. 2007). The first two are almost restricted to shallow areas, like bays, estuaries, and rocky shores, where wave and tide can have a prevalent effect on adult populations, and where most larvae start their journey after hatching (Fernandes et al. 2002; Bueno et al. 2010). These two factors, therefore, rule the transport of eggs, larvae, and other propagules on a small temporal (hours to days) and spatial (<10 km) scale. In contrast, wind-driven currents usually dominate after wave/tidal forces in both coastal and oceanic regions and are responsible for long-distance models of larval transport: offshore, shoreline, and onshore (Shanks 1995; Pineda 2000). According to the timing of the larval release, these complementary forces can have distinct, sometimes opposite, effects on benthic-pelagic dynamics. For example, the periwinkles *Nodilittorina lineolata* that inhabit the upper shore in southeast Brazil release their egg capsules coincident to ebb tides and onshore currents, suggesting retention close to the parental population (Bueno et al. 2010).

The strategy by which a larva develops into an adult may reflect the populational biogeographic distribution of a species and those with a broader distribution should spread their larvae over larger areas than those restricted or endemic. As an example, swimming crabs (Portunidae) living on sand beaches and estuaries can release millions of zoea larvae per female, which pass through numerous stages and long periods over the shelf. In contrast, shore crabs (Grapsidae) and spider crabs (Majidae) inhabiting rocky shores produce far less numerous offspring, which go through no more than four stages and can survive for shorter periods in coastal areas. This example can explain the distribution of shore crabs in isolated islands, like the St. Paul's Rocks in Brazil. *Grapsus grapsus* inhabit rocky shores at the oceanic St. Paul's Rocks and despite having the same behavior of larval release at nocturnal high tides as most coastal species, any larvae that succeeded to recruit must have quickly returned or even stayed nearby the settlement site (Koettker et al. 2010). This general model of benthic-pelagic coupling is widely spread over distinct groups, like barnacles, mussels, snails, corals, polychaetes, and cleaner shrimps, but also over ecosystems, like rocky shores, reefs, and estuaries. Surprisingly, the assumption that the broader the distribution, the larger the spread is reductionism and is not always true. Many species have larvae transported offshore and shoreline for over 100 km (Scheltema 1986; Cowen 2002), after which many return to the adult sites by onshore currents. In terms of connectivity, the final result may be very similar between return and retention if most replenishment is due to genetically related larvae and the dispersion is minimal, the so-called self-recruitment (Swearer et al. 2002). The implications of self-recruitment in the BPC are not well understood yet, but we may argue that at least some of the adult rocky shore populations in the Brazilian islands, like some fish populations living hundreds of kilometers apart from the continent, are not fully open (Quimbayo et al. 2019). In general, there exists a significant gene flow among isolated populations distributed over wide geographic regions that is due to larval transport and dispersion (Fogarty and Botsford 2006; Jones et al. 2009).

Settlement and Recruitment

Besides metamorphosis, benthic meroplanktonic organisms in rocky shores might deal with another crucial process in the life cycle—settlement—when a larva chooses the definitive place for its adult's entire life. Even though secondary displacement or post-settlement migration can occur for non-sessile organisms, like lobsters and polychaetes, it is restricted to proportionally short distances for most species.

The influence of settlement and recruitment upon BP coupling was only partially understood at the beginning of the 1980s. At that time, there was a consensus that younger stages like larvae were far more susceptible to shifts in environmental conditions than older ones (Thorson 1950), and most researchers assigned variability in recruitment exclusively to post-settlement events. Any subsequent process after settlement, like replenishment of adult stock, would be therefore dependent on the success in the recruitment phase. Following this concept of larval susceptibility to the environment, many researchers have tried to address the significance of recruitment to the community structure, even though most of them did not find any significant relationship between larval stock and recruitment. This unexpected poor predictive power highlighted the concept of the "planktonic mystery stage," and therefore many researchers focused on the "new" supply-side ecology (Underwood 1979; Lewin 1986; Underwood and Fairweather 1989). This new concept suggests that the recruitment of new individuals from the available larval pool should be somehow essential to the equilibrium of adult stock over time. The awareness of the lack of information about the mechanisms driving the supply of new individuals to the community fueled the supply-side ecology in the 1990s (Lewin 1986; Roughgarden et al. 1988).

According to Menge (1991), the influence of recruitment on the replenishment of the adult population increases relative to competition among juveniles and predation when the density of either settlers or recruits is low (no saturation condition). In contrast, when settlers' density is high enough to saturate recruitment, then competition, predation, and physical disturbance become significantly more important to the replenishment of the adults, weakening the stock-recruitment relationship. Perhaps the difficulty to access small recruits, mainly in soft bottoms like mud, led researchers to generalize the lack of causality that is most common in rocky shore communities, where larval pool tends to keep "saturate status" all over the year. Besides, recruitment success is often inferred weeks after actual settlement, not to mention that soft substrates are more widespread than hard ones (Fraschetti et al. 2003).

This generalization could explain the apparent lack of causality between recruitment rate and adult community structure under high larval density that usually ends with a high settlement rate. In contrast, the weak predictability in theoretical models does not reveal the actual mechanism by which adult assemblies are structured. For example, the replenishment of adult reef fish and rocky shore-associated populations seems sometimes unpredictable and episodic, mostly regulated by nonlinear processes (Dixon et al. 1999). Any attempt to model the benthic-pelagic coupling usually ends up with a sequence of events like "hatching," "transport," "retention," "settlement," and "recruitment." These processes are grouped relative to the metamorphosis as pre- and post-settlement events. Settlement is a biological event that takes a few hours to a few days, including sinking and searching for a favorable site, fixing or burrowing at the sediment, and finally metamorphosis (Fraschetti et al. 2003). Therefore, the larval settlement includes too many steps, and its use as a milestone in BP coupling is less informative than metamorphosis.

Relatively to settlement, recruitment is even less objective and an observer-defined concept (Booth and Brosnan 1995; Pineda et al. 2009) that starts with the definitive entry into the benthic population of individuals of some specific size, without already being an adult (Booth and Brosnan 1995). Regardless of the adopted nomenclature, most authors agree that the survival rate in each phase depends on intrinsic factors. Predation is one of the most important factors affecting community structure at all levels from pre- to post-settlement (Table 3.2). Differently, competition can be irrelevant to one phase but significant to the next. For instance, competition for space is irrelevant to larval survival in a vast 3D pelagic world but very relevant to settlement and post-settlement survival rates in a disputed 2D hard rocky shore. Thus, once a larva starts searching for a settlement place, it starts to compete with others for available space. Since not all larvae will find a proper place to settle, it is expected that only those more "competent" will succeed. So, competition for space tends to be highly relevant for hard-bottom post-settlement assemblies. Following this concept, the dispersion by unfavorable currents seems to be negligible to most lecithotrophic larvae, like the larvae of the sponge *Paraleucilla magna* in Guanabara Bay (Padua et al. 2013).

As previously mentioned, settlement and recruitment can be synchronous for some species but not necessarily at the same time for the entire community. The mismatch in time between taxonomic groups reflects the factors that drive synchronism, like larval behavior (Yoshinagua et al. 2010). Barnacles and mussels—two well-studied and dominant organisms on rocky shores—match their settlement to the spring tide, the wind-driven onshore currents, and the spring bloom of phytoplankton (Skinner and Coutinho 2002; Mazzuco et al. 2015, 2018; Fernandes et al. 2017).

A seasonal peak due to synchronism in the larval settlement suddenly shifts the recruitment-limited condition of the benthic community to a settlers-saturated

Table 3.2 Effectiveness (low, high) of major factors upon the primary and secondary recruitment (pre- and post-settlement assemblies) of planktotrophic/lecithotrophic (Pktroph/Lectroph), and hard/soft-bottom communities

	Pre-settlement		Post-settlement	
Factor	Pktroph	Lectroph	Hard	Soft
Competition for space	Low[a]	Low[a]	High	Low
Competition for food	High	–	High	Low[a]
Predation	High	High	High	High
Dispersion by unfavorable current	High	Low	Low	Low
Dispersion by self-displacement	Low	Low	Low[b]	High
Physical disturbance (wave, tide)	High	High	High	Low

[a]Competition tends to be highly variable according to the local community to allow any consistent generalization. Larvae are competitive for space during settlement when a larva must find the best place to metamorphose. Competition for food is high mostly in deep soft-bottom communities but can be irrelevant in shallow water, like mangroves, where organic matter is constantly provided by sinking POM

[b]Secondary dispersion is the most important factor affecting recruitment of surgeonfishes in Caribbean reefs (Robertson et al. 1988)

status, increasing the relevance of post-settlement competition and predation in regulating adult replenishment (Bruce 2000). Differently from rocky shores, in the Marapanim estuary, the coastal region of Amazon, the absence of synchronism in the settlement of megalopae of *Petrolisthes* (Decapoda) reveals recruitment as a constraint to the replenishment of adult stock (Oliveira et al. 2013). In contrast, under asynchronous recruitment, there is no peak in larval supply and the recruitment-stock relationship becomes more evident (Holm 1990).

The occurrence of favorable onshore currents is particularly relevant to increasing the settlement, generating strong synchronism under specific local conditions. In general, easterly trade winds on any west coast tend to promote offshore currents as well as upwelling events thus increasing larval dispersion to the shelf. The relaxing of upwelling is the favorable timing for the larvae to return to the coast and settle, as observed on the coast of Oregon (USA), Peru, and Chile (Roughgarden et al. 1988; Farrell et al. 1991; Navarrete et al. 2005). In contrast, the same trade wind on east coasts in the intertropical zone tends to create onshore currents and hence a pulse of settlement at fall (Johnson and Soltis 2017), while in temperate zones, settlement can be synchronized to upwelling (Menge et al. 2003; Wing et al. 2003).

On the Brazilian coast close to latitude 23°S, the wind-driven upwelling creates unfavorable conditions for settlement (Mazzuco et al. 2015), similar to the Chilean/ Peruvian coasts. Higher settlement rates in this region are strongly dependent upon the arrival of cold fronts (Skinner and Coutinho 2002; Bueno et al. 2010; Mazzuco et al. 2018), whose southwest winds weaken upwelling and promote onshore currents. Post-settlement processes, like predation and competition, become locally important after such pulses to recruit juveniles that will then replenish the adult population. Thus, oceanographic gradients related to offshore and onshore currents mediate the patterns of community structure by combining high settlement rates with food-web cascade effects (Menge et al. 2003).

Anthropogenic Interferences in Benthic-Pelagic Coupling

Fisheries and Whaling

Perhaps the most ancient, well known, and widespread human impact on the BPC comes from fisheries and overfishing. By removing, not always for food (fur, oil, entertainment), both pelagic—fish, squid, whales, and other less likely species such as penguins, sea turtles, dugongs, manatees, the now extinct Steller's sea cow—and demersal/benthic species—shrimps, crabs, and bottom-dwelling fish such as flounders and cod—humans have long been altering marine ecosystems, by selective removal of species either from the water column or from the sea floor. As some authors put it (Pauly et al. 1998), we have been "fishing down the marine food web," that is, fishing for large predatory fish to the point of overexploitation and then moving down the food web. This led to widespread overfishing (FAO estimates that over

60% of the world's fisheries species are currently either overfished or on the verge thereof), sometimes beyond recovery, of important links in the marine food web. Removal through fisheries of pelagic predators has also been blamed for causing the worldwide decline of coral reefs, or "phase shifts" in benthic ecosystems, such as rocky shores and coral reefs, from coral-dominated (or other invertebrates) to algal-dominated ecosystems and from these, into diversity-impoverished sea-urchin barrens or even bare rock (e.g., Bellwood et al. 2004).

The removal of keystone pelagic predators seems to have deeply affected BPC and thus marine ecosystems, as some well-documented examples have shown. One of the best documented of these keystone predators is the sea otter (*Enhydra lutris*), overhunted for its fur along the Pacific coast of the USA, from California to Alaska. After recovering from near local extinction in Alaska, sea otter populations then declined again in western Alaska (Estes et al. 1998), apparently due to increased killer whale predation. Devoid of its natural predator, sea urchins increased in density, causing deforestation of kelp beds in the nearshore community. Although the causes remain unclear, anthropogenic changes in the offshore oceanic ecosystem may have triggered this chain of events, reducing kelp forests and the myriad of species that depend on them, including rocky shore and even terrestrial species (Estes et al. 1998). The phase shift from kelp forests to deforested sea urchin barrens was caused by the removal of a single species. Apparently, the increased predation by transient killer whales led not only to a decline in sea otter populations but also in pinnipeds such as seals and sea lions in southwest Alaska (Estes et al. 2009). One possible explanation (that Estes et al. deem unlikely) would be that killer whales, which fed extensively on great whales, expanded their diets to include a higher proportion of sea otters and pinnipeds after a sharp reduction in great whale numbers from post-World War II industrial whaling. If overfishing is at the root of these changes, the implications for fisheries management are profound. Thus, changes in the open ocean appear to have driven killer whales to prey on sea otters, affecting the entire structure of the coastal ecosystem (Estes et al. 1998, 2009).

Despite the official worldwide stale on whaling in the world since the late 1980s, some countries (Japan, Norway, Denmark, Iceland, Canada, the USA, Russia, and South Korea) still follow a more individualist approach to whale conservation and, in some cases, under the guise of food needs, entertainment industry (display and "shows" in aquaria), or scientific research, keep hunting hundreds to thousands of whales and dolphins every year. There is an obvious consensus among scientists on the detrimental effects of the removal of these live, long-living, massive individuals from the marine food web and the pelagic environment. Less well known and discussed are, perhaps, the effects of whaling on depriving the deep-sea ecosystems of a major nutritional source: whale carcasses.

Obviously, deep-sea ecology lags behind the study of other marine environments such as rocky shores. This, however, is not due to a lack of interest but a consequence of the immense technological hurdles posed by the deep-sea environment, such as hydrostatic pressure. Today only a few submersibles are capable of manned deep-sea diving, like the Shinkai 6500. At the base of the São Paulo Ridge, 700 km off the Brazilian coast, scientists onboard this submersible met a thriving community living off a few left vertebrae of a long-gone (ca. 5–10 years) Antarctic minke whale resting on a thin sediment layer on top of the basaltic rock bed at 4200 meters depth (Sumida et al. 2016). This is the deepest natural whale carcass ever found, and the ongoing study of its fauna is shedding light on benthic-pelagic coupling in the deep sea, and even in the connectivity between distant ocean basins.

Whale carcasses are considered the largest organic inputs reaching the deep-sea benthic environment in a single event, and they attract a plethora of both opportunistic and specialist organisms. Specialized organisms, such as bone-eating polychaetes of the genus *Osedax*, evolved in these habitats for millions of years. Whale carcass falls, as well as chemosynthetic-based communities in hydrothermal vents and cold seeps, constitute true oases of evolutionary novelty and biodiversity in the deep sea, and all these deep-sea ecosystems, however ephemeral, seem to be highly interconnected. Whale-fall communities undergo several successional stages, as initially necrophages/scavengers remove the soft tissues, while opportunists colonize the bones and the surrounding sedimentary seabed. A "sulphophilic" stage starts when an anaerobic microbial breakdown of the then organically rich sediments, as well as the lipid-rich skeleton creates elevated flows of reduced compounds, thus allowing the development of a chemosynthesis-based community. The fauna at this stage overlaps that of hydrothermal vents, cold seeps, and woodfalls. Whale falls can thus act as "stepping stones" for deep-sea fauna dispersal and evolution (Smith et al. 2015).

More likely to occur worldwide along whale migration routes, breeding, and feeding areas, natural whale-fall communities seem to connect distant oceanic basins, such as the SW Atlantic and the NE Pacific. Although apparently the majority of the at least 41 species sampled from this whale fall are new to science (Sumida et al. 2016), they belong to genera previously recorded in hydrothermal vents, cold seeps, and wood and whale falls found in the NE Pacific, and some species appear to have even a worldwide distribution (e.g., *Osedax rubiplumus*), suggesting indeed the existence of a large deep-sea, inter-basin connectivity. Apparently, a high (pre-whaling) abundance of large marine mammals has formed a reliable enough organic source to develop a distinct "following" in the deep sea of organisms adapted entirely to subsistence in carcasses (Tunnicliffe et al. 2003). As these communities have, on an evolutionary time scale, depended on whale carcass falls to disperse and evolve, what was then the effect of historical whale hunting in deep-sea communities? As many "stepping stones" never formed in the first place, did we break the connectivity between ocean basins by hunting whales for their meat and oil? Are we, by allowing the continuation of whaling, presently destroying a long-evolving deep-sea connectivity? How many yet unknown marine species depend on natural whale falls? What are the effects of preventing the "stepping stones" for deep-sea fauna to even form? Are there other, yet undiscovered, deep-sea oases in the SW Atlantic (e.g., somewhere along the southern portion mid-Atlantic Ridge), supplying or exchanging the diversity of marine life detected in this whale-fall ecosystem? These are all questions we can not presently answer precisely, but that suggests these delicate and ephemeral benthic deep-sea ecosystems are deeply interconnected and interdependent through the BP pathway.

Pollution

There are several ways by which pollutants may affect benthic-pelagic coupling (BPC). Many persistent organic pollutants (POPs) derived from human industrial activities are spilled in the pelagic environment and returned to us as contaminants in benthic species as seafood. Increased attention is given today to the pollution by microplastics, which accumulate in benthic filter-feeders, such as brown mussels (Birnstiel et al. 2019), and escalate all the way through the food web into our tables. However, it is unclear how these pollutants interfere directly with the benthic-pelagic coupling (BPC). We will therefore focus on a few examples in which the consequences to BPC are more striking, such as organotin pollution from antifouling (AF) paints and cultural eutrophication.

Organotins from Antifouling Paints

These compounds provided a more efficient antifouling (AF) agent for coatings designed to inhibit the attachment of hard-bottom marine benthic organisms on man-made surfaces such as ship hulls. In other words, organotins were more efficient biocides in AF paints for preventing antifouling growth than the previous generation of copper-based AF paints. Organotins such as trybutiltin (TBT) were thus widely used until the worldwide ban by IMO (International Maritime Organization) signatory countries effective from 2008 on. In Brazil, NORMAM 23, enforced by the Brazilian Navy in 2007, banned the use of TBT in AF paints along the Brazilian coast. The bans followed international awareness about the toxicity, persistence, and undesired effects of organotins in non-target marine species, such as dogwhelks (gastropod mollusks of the *Nucella* genus). This class of chemicals acts as an endocrine disruptor in many species, including gastropod mollusks of several genera. Females of these gastropods, when exposed to water contaminated by TBT, develop non-functional masculine sexual organs in several degrees (pseudo-hermaphroditism or imposex). Therefore, imposex in female gastropods impairs reproduction of affected individuals and is a biomarker of TBT contamination. Since 2008, a series of studies have demonstrated the recovery of contaminated ports and marinas such as in Portugal, England, Alaska, and even the nearby Guanabara Bay (Fernandez et al. 2005). Since the ban on the Brazilian coast was effective in 2007, Toste et al. (2011) investigated the levels of imposex in the local whelk *Stramonita haemastoma* (an edible species locally known as "corogondó") before (2001) and after the ban (2008) in the sensitive coast of Arraial do Cabo (Rio de Janeiro State, southeastern Brazil). The area has a port and an adjacent marina and is one of the only regions of the SW Atlantic where upwelling occurs, therefore being of important significance for regional fisheries. Such importance led to the area becoming a Marine Extractive Reserve in 1997, allowing traditional, local fisheries to continue but being otherwise a Marine Protected Area (MPA). Surprisingly, Toste et al. (2011)

found evidence that, instead of decreasing, TBT pollution as a percentage of imposex in *S. haemastoma* females increased in intensity in some areas (such as Maramutá, in the inner side of Cabo Frio Island, relatively pristine), but led to the almost complete absence of this rocky shore predator in the area near Forno port (Anjos beach) and adjacent areas after the ban (2008), in comparison with the same 9 stations monitored in 2001. This disappearance is surprising (other port areas can have up to 100% imposex, therefore depending completely on the supply of gastropod larvae from unaffected areas) and suggests that the rocky shore community here may be disrupted. Mussels are the main prey item of *Stramonita*, and, although speculative, it is not unlikely that this area may have reduced diversity due to space monopolization by mussels, or even be more susceptible to bioinvasion (i.e., have increased invasibility) once a benthic predator is removed. In fact, the exotic vermetid *Eualetes tulipa* was recently detected in the port area of Anjos beach (Spotorno-Oliveira et al. 2017), where the piers and rock jetty have already been massively colonized by invasive sun corals (*Tubastraea* spp.).

Cultural Eutrophication

Large-scale mechanized agriculture and urbanization usually result in nutrient enrichment in rivers and, ultimately, lead to coastal eutrophication. The large supply of extra nutrients from agricultural and urban runoff, including nitrogen (N) and the usually limiting phosphorus (P), leads to more frequent and massive phytoplankton blooms, including harmful algal blooms—HABs (Diaz and Rosenberg 2008). These blooms grow exponentially until massive die-offs occur when conditions change or nutrients exhaust, greatly accelerating the otherwise natural transfer of organic matter from the pelagic to the benthic environment. This excess of organic matter covers the sea bottom and creates perfect conditions for hypoxic or anoxic, acid-producing (H_2S) degradation of the organic matter surplus, creating large dead zones where only a few species can survive (Diaz and Rosenberg 2008).

Considered one of the world's largest, the Gulf of Mexico dead zone, off the coast of Louisiana, is an annually recurring event and occurs every summer. NOAA estimated that ca. 156,000 t of N and 25,300 t of P were discharged by the Mississipi river into the Gulf of Mexico in May 2019, a higher than average nutrient input that can cause a major dead zone in summer. The annual discharge of N and P is estimated at an average of 1.7 million tons. An estimated number of ca. 400 dead zones occur around the globe, some as small as 1 km^2 (0.4 mi^2), while others can reach 70,000 km^2 (27,000 mi^2). Climate has a significant impact on the growth and decline of dead zones. During spring, as rainfall increases, nutrient-enriched waters flow down the Mississippi River, at the same time as sunlight increases. Algal growth in the dead zones thus increases dramatically. In fall, tropical storms enter the Gulf of Mexico and break up the dead zones. The cycle then restarts every spring. An increase in N and P generally cause Cyanobacteria (blue-green algae) to bloom. Other algae are usually consumed by zooplankton and thus do not accumulate to the same extent. As many blue-greens do not provide good food for zooplankton and fish, they accumulate in the water column, die, and then decompose. Bacterial degradation consumes the oxygen in the water, creating hypoxia. Dead zones, however, are not always anthropogenic—natural factors such as coastal upwelling can also lead to the formation of dead zones. However, the widespread use of chemical fertilizers is considered the major human-related cause of dead zones worldwide. Runoff from sewage and urban land use also contribute to the formation of dead zones. Ironically, meat-eating can reduce seafood availability, as most of the crops grown in the Mississippi River basin are used as major feed components in the production of meat animals for agribusiness.

Due to the hypoxic conditions, marine life within these areas tends to be very scarce. Most fish and other vagile (motile) organisms tend to move out of dead zones as dissolved oxygen (DO) levels fall, but benthic populations suffer heavy losses when DO reaches 0.5 $mg.l^{-1}$. Under severe anoxic conditions, shifts in microbial life also occur, with increased abundance of anaerobic bacteria, which use nitrate, sulfate, or iron reduction as energy sources. Sulfur reduction is particularly harmful, as hydrogen sulfide (H_2S) is acidic and toxic, increasing the risk of further benthic mortality. Recovery of benthic communities depends upon the composition and larval transport from adjacent, unharmed areas for recruitment (a similar situation appears to happen in bays, marinas and other areas heavily polluted by organotins from AF paints, particularly to species vulnerable to endocrine disruption, such as *Stramonita haemastoma* in the Brazilian coast). This usually results in a community shift toward opportunistic species, potentially disrupting historic benthic compositions. These effects lower the diversity of yield in fisheries, but can lead to temporary increases in fisheries of species such as anchovies. However, the overall net result is considered a great loss in both fisheries and tourist revenue (Diaz and Rosenberg 2008).

References

Amieva MR, Reed CG (1987) Functional morphology of the larval tentacles of Phragmatopoma californica (Polychaeta: Sabellariidae): composite larval and adult organs of multifunctional significance. Mar Biol 95:243–258. https://doi.org/10.1007/BF00409012

Anger K (2002) The biology of decapod Crustacean larvae. J Exp Mar Bio Ecol 14:89–90

Barletta M, Barletta-Bergan A (2009) Endogenous activity rhythms of larval fish assemblages in a mangrove-fringed estuary in North Brazil. Open Fish Sci J 2:15–24. https://doi.org/10.2174/1874401x00902010015

Barnes H, Barnes M (1965) Egg size, Nauplius size, and their variation with local, geographical, and specific factors in some common cirripedes. J Anim Ecol, 34(2):391–402. https://doi.org/10.2307/2656

Bashevkin SM, Christy JH, Morgan SG (2020) Costs and compensation in zooplankton pigmentation under countervailing threats of ultraviolet radiation and predation. Oecologia 193:111–123. https://doi.org/10.1007/s00442-020-04648-2

Bellwood DR, Hughes TP, Folke C, Nyström M (2004) Confronting the coral reef crisis. Nature 429:827–833. https://doi.org/10.1038/nature02691

Birnstiel S, Soares-Gomes A, da Gama BAP (2019) Depuration reduces microplastic content in wild and farmed mussels. Marine Pollution Bulletin 140:241–247. https://doi.org/10.1016/j.marpolbul.2019.01.044

Booth DJ, Brosnan DM (1995) The role of recruitment dynamics in rocky shore and coral reef fish communities. Adv Ecol Res 26:309–385

Bruce M (2000) Top-down and bottom-up community regulation in marine rocky intertidal habitats. J Exp Mar Bio Ecol 250:1–26

Bueno M, Flores AAV (2010) Tidal-amplitude rhythms of larval release: variable departure from presumed optimal timing among populations of the mottled shore crab. J Mar Biol Assoc UK 90:859–865. https://doi.org/10.1017/S0025315410000044

Bueno M, Moser GAO, Tocci BRC, Flores AAV (2010) Retention-favorable timing of propagule release in barnacles and periwinkles. Mar Ecol Prog Ser 414:155–165. https://doi.org/10.3354/meps08715

Coutinho R, Yaginuma LE, Siviero F, dos Santos JCQP, López MS, Christofoletti RA, Berchez F, Ghilardi-Lopes NP, Ferreira CEL, Gonçalves JEA, Masi BP, Correia MD, Sovierzoski HH, Skinner LF, Zalmon IR (2016) Studies on benthic communities of rocky shores on the Brazilian coast and climate change monitoring: status of knowledge and challenges. Braz J Oceanogr 64:27–36

Cowen RK (2002) Larval dispersal and retention and consequences for population connectivity. In: Sale PF (ed) Coral reef fishes: dynamics and diversity in a complex ecosystem. ScienceDirect, pp 149–170

Derenbach JB, Gereck MV (1980) Interference of petroleum hydrocarbons with the sex pheromone reaction of Fucus vesiculosus (L.). J Exp Mar Biol Ecol 44:61–65. https://doi.org/10.1016/0022-0981(80)90101-X

Diaz RJ, Rosenberg R (2008) Spreading dead zones and consequences for marine ecosystems. Science 321(5891):926–929. https://doi.org/10.1126/science.1156401

Estes JA, Tinker MT, Williams TM, Doak DF (1998) Killer whale predation on sea otters: linking oceanic and nearshore ecosystems. Science 282:472–476. https://doi.org/10.1126/science.282.5388.473

Estes JA, Doak DF, Springer AM, Williams TM (2009) Causes and consequences of marine mammal population declines in southwest Alaska: a food-web perspective. Philos. Trans. R. Soc. Lond. B Biol. Sci. 364:1647–1658. https://doi.org/10.1098/rstb.2008.0231

Farrell TM, Bracher D, Roughgarden J (1991) Cross-shelf transport causes recruitment populations in central California to intertidal. Limnol an Oceanogr 36:279–288

Fernandes LDA, Bonecker SLC, Valentin JL (2002) Dynamic of decapod Crustacean larvae on the entrance of Guanabara Bay. Braz Arch Biol Technol 45:491–498

Fernandes LDA, Netto EBF, Coutinho R (2017) Inter-annual cascade effect on marine food web: a benthic pathway lagging nutrient supply to pelagic fish stock. PLoS One 12:e0184512. https://doi.org/10.1371/journal.pone.0184512

Fraschetti S, Giangrande A, Terlizzi A, Boero F (2003) Pre- and post-settlement events in benthic community dynamics. oceanol Acta 25:285–295

Fogarty MJ, Botsford LW (2006) Metapopulation dynamics of coastal decapods. In: Kritzer JP, Sale PF (eds) Marine metapopulations. Elsevier, pp 271–319

de Melo Júnior M, Melo PAM de C, Paranaguá MN, Neumann-Leitão S, Schwamborn R (2016) Composition of decapod larvae in a northeastern Brazilian estuarine inlet over a full tidal cycle. Lat Am J Aquat Res 44:401–410. https://doi.org/10.3856/vol44-issue2-fulltext-21

de Oliveira DB, Silva DC, Martinelli-Lemos JM (2013) Larval and adult density of the porcellanid crab petrolisthes armatus (Anomura: Porcellanidae) in an Amazon estuary, Northern Brazil. Zoologia 30:592–600. https://doi.org/10.1590/S1984-46702013005000002

Diele K, Simith DJB (2006) Salinity tolerance of Northern Brazilian mangrove crab larvae, Ucides cordatus (Ocypodidae): necessity for larval export? Estuar Coast Shelf Sci 68:600–608. https://doi.org/10.1016/j.ecss.2006.03.012

Dixon PA, Milicich MJ, Sugihara G (1999) Episodic fluctuations in larval supply. Science (80–) 283:1528–1530

Dyachuk V, Odintsova N (2009) Development of the larval muscle system in the mussel Mytilus trossulus (Mollusca, Bivalvia). Dev Growth Differ 51:69–79. https://doi.org/10.1111/j.1440-169X.2008.01081.x

Emlet RB (1986) Facultative Planktotrophy in the Tropical Echinoid Clypeaster rosaceus (Linnaeus) and a comparison with obligate planktotrophy in Clypeaster subdepressus (Gray) (Clypeasteroida: Echinoidea). J Exp Mar Bio Ecol 95:183–202

Gallardo CS, Penchaszadeh PE (2001) Hatching mode and latitude in marine gastropods: Revisiting Thorson's paradigm in the southern hemisphere. Mar Biol 138:547–552. https://doi.org/10.1007/s002270000477

Gallego A, North EW, Houde ED (2012) Understanding and quantifying mortality in pelagic, early life stages of marine organisms – old challenges and new perspectives. J Mar Syst 93:1–3. https://doi.org/10.1016/j.jmarsys.2011.10.012

Griffiths JR, Kadin M, Nascimento FJA, Tamelander T, Törnroos A, Bonaglia S, Bonsdorff E, Brüchert V, Gårdmark A, Järnström M, Kotta J, Lindegren M, Nordström MC, Norkko A, Olsson J, Weigel B, Žydelis R, Blenckner T, Niiranen S, Winder M (2017) The importance of benthic–pelagic coupling for marine ecosystem functioning in a changing world. Glob Chang Biol 23:2179–2196. https://doi.org/10.1111/gcb.13642

Hansson LA (2000) Induced pigmentation in zooplankton: a trade-off between threats from predation and ultraviolet radiation. Proc Biol Sci. 267:2327–2331. https://doi.org/10.1098/rspb.2000.1287

Holm ER (1990) Effects of density-dependent mortality on the relationship between recruitment and larval settlement. Mar Ecol Prog Ser 60:141–146

Ims RA (1990) The ecology and evolution of reproductive synchrony. Trends Ecol Evol 5:135–140. https://doi.org/10.1016/0169-5347(90)90218-3

Johnson KB, Soltis KA (2017) Larval supply and recruitment of the ivory barnacle Balanus eburneus (Gould, 1841) on oyster cultch used in oyster-reef restoration. J Crustac Biol 37:243–248. https://doi.org/10.1093/jcbiol/rux015

Jones GP, Almany GR, Sale P, Van OM (2009) Larval retention and connectivity among populations of corals and reef fishes: history, advances and challenges. Coral Reefs 28:307–325. https://doi.org/10.1007/s00338-009-0469-9

Junqueira AOR, Ventura CRR, Carvalho ALPS, Schmidt AJ (1997) Population recovery of the sea urchin Lytechinus variegatus in a seagrass flat (Araruama Lagoon, Brazil): the role of recruitment in a disturbed environment. Invert Reprod Dev 31:143–150. https://doi.org/10.1080/07924259.1997.9672572

Koettker AG, Freire AS, Sumida PYG (2010) Temporal, diel and spatial variability of decapod larvae from St Paul's Rocks, an equatorial oceanic island of Brazil. J Mar Biol Assoc UK 90:1227–1239. https://doi.org/10.1017/S0025315409990890

Levin LA, Creed EL (1986) Effect of temperature and food availability on reproductive responses of Streblospio benedicti (Polychaeta: Spionidae) with planktotrophic or lecithotrophic development L. Mar Biol 92:103–113

Levin LA, Caswell H, DePatra KD, Creed EL (1987) Demographic consequences of larval development mode: planktotrophy vs. lecithotrophy in Streblospio benedicti. Ecology 68:1877–1886

Lewin R (1986) Supply-side ecology. Science (80–) 234:25

López MS, Coutinho R (2008) Acoplamento Plâncton-Bentos: O Papel Do Suprimento Larval Na Estrutura Das Comunidades Bentônicas De Costões Rochosos. Oecologia Aust 12:575–601. https://doi.org/10.4257/oeco.2008.1204.01

Mazzuco ACA, Christofoletti RA, Pineda J, Starczak VR, Ciotti AM (2015) Temporal variation in intertidal community recruitment and its relationships to physical forcings, chlorophyll-a concentration and sea surface temperature. Mar Biol 162:1705–1725. https://doi.org/10.1007/s00227-015-2689-6

Mazzuco AC de A, Christofoletti RA, Coutinho R, Ciotti ÁM (2018) The influence of atmospheric cold fronts on larval supply and settlement of intertidal invertebrates: case studies in the Cabo Frio coastal upwelling system (SE Brazil). J Sea Res 137:47–56. https://doi.org/10.1016/j.seares.2018.02.010

McClintock JB, Swenson DP, Steinberg DK, Michaels AA (1996) Feeding-deterrent properties of common oceanic holoplankton from Bermudian waters. Limnol Oceanogr Oceanogr 41:798–801

McEdward LR, Janies DA (1993) Life cycle evolution of asteroids: what is a larva? Biol Bull 184:255–268. https://doi.org/10.2307/1542444

McManus MA, Woodson CB (2012) Plankton distribution and ocean dispersal. J Exp Mar Bio Ecol 215:1008–1016. https://doi.org/10.1242/jeb.059014

Meirelles CAO, Matthews-Cascon H (2005) Spawn and larval development of Pleuroploca aurantiaca (Lamarck, 1816) (Gastropoda: Fasciolariidae) from northeast Brazil. Sci Mar 69:199–204

Menge BA (1991) Relative importance of recruitment and other causes of variation in rocky intertidal community structure. J Exp Biol 146:69–100

Menge BA, Lubchenco J, Bracken MES, Chan F, Foley MM, Freidenburg TL, Gaines SD, Hudson G, Krenz C, Leslie H, Menge DNL, Russell R, Webster MS (2003) Coastal oceanography sets the pace of rocky intertidal community dynamics. Proc Natl Acad Sci U S A 100:12229–12234. https://doi.org/10.1073/pnas.1534875100

Mikhlina A, Ekimova I, Vortsepneva E (2020) Functional morphology and post-larval development of the buccal complex in Eubranchus rupium (Nudibranchia: Aeolidida: Fionidae). Zoology 143:1–16. https://doi.org/10.1016/j.zool.2020.125850

Mileikovsky SA (1971) Types of larval development in marine bottom invertebrates, their distribution and ecological significance: a re-evaluation. Mar Biol 10:193–213. https://doi.org/10.1007/BF00352809

Morgan SG (1992) Predation by planktonic and benthic invertebrates on larvae of estuarine crabs. J Exp Mar Bio Ecol 163:91–110. https://doi.org/10.1016/0022-0981(92)90149-5

Morgan SG, Christy JH (1997) Planktivorous fishes as selective agents for reproductive synchrony. J Exp Mar Biol Ecol 209:89–101

Navarrete SA, Wieters EA, Broitman BR, Castilla JC (2005) Scales of benthic–pelagic coupling and the intensity of species interactions: from recruitment limitation to top-down control. Proc Natl Acad Sci 102:18046–18051

Padua A, Lanna E, Zilberberg C, de Paiva PC, Klautau M (2013) Recruitment, habitat selection and larval photoresponse of Paraleucilla magna (Porifera, Calcarea) in Rio de Janeiro, Brazil. Mar Ecol 34:56–61. https://doi.org/10.1111/j.1439-0485.2012.00524.x

Pauly D, Christensen V, Dalsgaard J, Froese R, Torres F (1998) Fishing down marine food webs. Science 279:860–863

Pechenik JA (1999) On the advantages and disadvantages of larval stages in benthic marine invertebrate life cycles. Mar Ecol Prog Ser 177:269–297

Pepe-Victoriano R, Miranda L, Ortega A, Merino GE (2021) Descriptive morphology and allometric growth of the larval development of Sarda chiliensis chiliensis (Cuvier, 1832) in a hatchery in northern Chile. Aqua Rep 19:11. https://doi.org/10.1016/j.aqrep.2020.100576

Pineda J (2000) Linking larval settlement to larval transport: assumptions, potentials, and pitfalls. Oceanogr East Pacific 1:84–105

Pineda J, Hare J, Sponaugle S (2007) Larval transport and dispersal in the coastal ocean and consequences for population connectivity. Oceanography 20:22–39. https://doi.org/10.5670/oceanog.2007.27

Pineda J, Reyns NB, Starczak VR (2009) Complexity and simplification in understanding recruitment in benthic populations. Popul Ecol 51:17–32

Quimbayo JP, Dias MS, Kulbicki M, Mendes TC, Lamb RW, Johnson AF, Aburto-Oropeza O, Alvarado JJ, Bocos AA, Ferreira CEL, Garcia E, Luiz OJ, Mascareñas-Osorio I, Pinheiro HT, Rodriguez-Zaragoza F, Salas E, Zapata FA, Floeter SR (2019) Determinants of reef fish assemblages in tropical Oceanic islands. Ecography (Cop) 42:77–87. https://doi.org/10.1111/ecog.03506

Robertson DR, Green DG, Victor BC (1988) Temporal coupling of production and recruitment of larvae of a Caribbean Reef Fish. Ecology 69:370. https://doi.org/10.2307/1940435
Roughgarden J, Gaines SD, Possingham H (1988) Recruitment dynamics in complex life cycles. Science (80–) 241:1460–1466. https://doi.org/10.1126/science.11538249
Satuito CG, Natoyama K, Yamazaki M, Fusetani N (1994) Larval development of the mussel Mytilus edulis galloprovincialis cultured under laboratory conditions. Fish Sci 60:65–68. https://doi.org/10.2331/fishsci.60.65
Scheltema RS (1986) On dispersal and planktonic larvae of benthic invertebrates: an ecletic overview and summary of problems. Bull Mar Sci 39:290–322
Shanks AL (1995) Mechanisms of cross-shelf dispersal of larval invertebrates and fish. In: MacEdward LR (ed) Ecology of marine invertebrate larvae. CRC Press, London, pp 323–367
Skinner LF, Coutinho R (2002) Preliminary results on settlement of the barnacles Tetraclita stalactifera and Chthamalus bisinuatus on a Brazilian tropical rocky shore under upwelling conditions. Invertebr Reprod Dev 41:151–156. https://doi.org/10.1080/07924259.2002.9652746
Smith DL, Johnson KB (1996) A guide to marine coastal plankton and marine invertebrate larvae. Kendall/Hunt Pub. Co., Dubuque
Smith CR, Glover AG, Treude T, Higgs ND, Amon DJ (2015) Whale-fall ecosystems: recent insights into ecology, paleoecology, and evolution. Ann Rev Mar Sci 7:571–596. https://doi.org/10.1146/annurev-marine-010213-135144
Sousa AM, Machado I, Pereira MO (2011) Phenotypic switching: an opportunity to bacteria thrive. In: Méndez-Vilas A (ed) Science against microbial pathogens: communicating current research and technological advances. Formatex Research Center, pp 252–262
Spotorno-Oliveira P, Coutinho R, Tâmega FT, de S (2017) Recent introduction of non-indigenous vermetid species (Mollusca, Vermeditae) to the Brazilian coast. Mar Biodivers 48:1931–1941. https://doi.org/10.1007/s12526-017-0702-7
Stanley JA, Hesse J, Hinojosa IA, Jeffs AG (2015) Inducers of settlement and moulting in post-larval spiny lobster. Oecologia 178:685–697. https://doi.org/10.1007/s00442-015-3251-4
Strathmann RR, Grunbaum D (2006) Good eaters, poor swimmers: compromises in larval form. Integr Comp Biol 46:312–322. https://doi.org/10.1093/icb/icj031
Sumida PYG, Alfaro-Lucas JMA, Shimabukuro M, Kitazato H, Perez JAA, Soares-Gomes A, Toyofuku T, Lima AOS, Ara K, Fujiwara Y (2016) Deep-sea whale fall fauna from the Atlantic resembles that of the Pacific Ocean. Sci Rep 6:1–9. https://doi.org/10.1038/srep22139
Swearer SE, Shima JS, Hellberg ME, Thorrold SR, Jones GP, Robertson DR, Morgan SG, Selkoe KA, Ruiz GM, Warner RR (2002) Evidence of self-recruitment in demersal marine populations. Bull Mar Sci 70:251–271
Tapia FJ, Pineda J (2007) Stage-specific distribution of barnacle larvae in nearshore waters: potential for limited dispersal and high mortality rates. Mar Ecol Prog Ser 342:177–190
Thompson JV (1828) On the metamorphoses of the Crustacea, and on the zoea, exposing their singular structure, and demonstrating that they are not, as has been supposed, a peculiar genus, but the larva of Crustacea. In: Zoological Researchers; or Natural History of nondiscript or imperfectly known animals, in a series of memoirs. Memoir I, pp 1–11
Thorson G (1950) Reproductive and larval ecology of marine bottom invertebrates. Biol Rev Camb Philos Soc 25:1–45
Todd CD (1979) Reproductive energetics of two species of dorid nudibranchs with planktotrophic and lecithotrophic larval strategies. Mar Biol 53:57–68
Toste R, Fernandez MA, Pessoa I, de A, Parahyba MA, Dore MP (2011) Organotin pollution at Arraial do Cabo, Rio de Janeiro State, Brazil: increasing levels after the TBT ban. Brazilian J Oceanogr 59(1):111–117
Tunnicliffe S, Juniper K, Sibuet M (2003) Reducing environments of the deep-sea floor. In: Tyler PA (ed.) Ecosystems of the deep oceans. Ecosystems of the World 28, Elsevier, Amsterdam, 81–110
Underwood A (1979) Ecology of intertidal gastropods. Adv Mar Biol 16:111–210

Underwood A, Fairweather P (1989) Supply-side ecology and benthic marine assemblages. Tree 4:16–20
Vance RR (1973) On reproductive strategies in marine benthic invertebrates. Am Nat 107:339–352
Vaughn D, Allen JD (2010) The peril of the plankton. Integr Comp Biol 50:552–570. https://doi.org/10.1093/icb/icq037
Ventura CRR, Falcão APC, Santos JS, Fiori CS (1997) Reproductive cycle and feeding periodicity in the starfish Astropecten brasiliensis in the Cabo Frio upwelling ecosystem (Brazil). Invert Reprod Dev 31:135–141. https://doi.org/10.1080/07924259.1997.9672571
Villwock JA, Lessa GC, Suguio K, Angulo RJ, Dillenburg SR (2005) Geologia e Geomorfologia de Regiões Costeiras. In: de Souza CRG, Suguio K, dos Oliveira AMS, de Oliveira PE (eds) Quaternário do Brasil. Holos, São Paulo, pp 94–113
Vuichard GS, Farías N, Luppi T (2013) Hatching and larval export of the intertidal crab Neohelice granulata in Mar Chiquita coastal lagoon, Argentina. Iheringia Série Zool 103:124–133. https://doi.org/10.1590/s0073-47212013000200007
Wilson JR, Harrison PL (1998) Settlement-competency periods of larvae of three species of scleractinian corals. Mar Biol 131:339–345. https://doi.org/10.1007/s002270050327
Wing SR, Botsford LW, Morgan LE, Diehk JM, Jose LC (2003) Inter-annual variability in larval supply to populations of three invertebrate taxa in the northern California current. Estuar Coast Shelf Sci 57:859–872
Yoshinagua MY, Sumida PYG, Silveira ICA, Ciotti AM, Gaeta SA, Pacheco LFCM, Koettker AG (2010) Vertical distribution of benthic invertebrate larvae during an upwelling event along a transect off the tropical Brazilian continental margin. J Mar Syst 79:124–133. https://doi.org/10.1016/j.jmarsys.2009.07.007

Chapter 4
Regulatory Processes of Biological Zonation on Rocky Shores

Ilana Rosental Zalmon, Frederico T. S. Tâmega, and Ricardo Coutinho

Abstract Intertidal rocky shores present several organisms that are commonly distributed in specific zones or particular associations parallel to the coast, with species variations over short horizontal distances. The vertical distribution is related mainly to a gradient of emersion/desiccation and also to others physical and biological factors. Benthic community zonation patterns are well known, and this subject has been discussed by several authors. For marine biologists, the zones are extremely useful in a practical sense because they provide a convenient descriptive framework into which specific observations can be fitted. This chapter intends to deepen into the post-recruitment interaction processes between benthic species (competition, predation, herbivory) and how this leads to zonation and community dynamics.

Introduction

Among coastal environments, rocky shores (Fig. 4.1a, b) is highlighted for their high biodiversity of organisms (Fig. 4.1c, d) including seaweeds, invertebrates, fishes, seabirds, in addition to their great ecological and economic importance (Horn et al. 1999; Harley et al. 2012) and the provision of essential environmental services (Hershner and Mason 2009).

Intertidal rocky shores present several organisms that are commonly distributed in specific zones or particular associations parallel to the coast, with species variations over short horizontal distances (Fig. 4.1e, f). The vertical distribution of these bands is related mainly to a gradient of emersion/desiccation and also to other physical and biological factors, including wave exposure, tidal dynamics (Denny and

I. R. Zalmon (✉)
Laboratório de Ciências Ambientais, Universidade Estadual do Norte Fluminense Darcy Ribeiro, Campos dos Goytacazes, RJ, Brazil
e-mail: ilana@uenf.br

F. T. S. Tâmega · R. Coutinho
Instituto de Estudos do Mar Almirante Paulo Moreira, Departamento de Biotecnologia Marinha, Arraial do Cabo, RJ, Brazil

R. Coutinho, R. A. Christofoletti (eds.), *Brazilian Rocky Shores*, Brazilian Marine Biodiversity, https://doi.org/10.1007/978-3-031-67206-4_4

Fig. 4.1 Examples of rocky shore environment (**a**, **b**). Biodiversity in the intertidal zone (*Ulva lactuca*, geniculate coralline algae, barnacles, and mollusks) (**c**). Biodiversity in the subtidal zone (non-geniculate and geniculate coralline algae, sea-urchins, and mollusks) (**d**). Intertidal rocky shore zonation (**e**, **f**). Species interaction as grazing (**g**) and competition for space (**h**). Scale bar: **c** (5 cm), **d** (10 cm), **e** (5 cm), **f** (5 cm), **g** (10 cm), **h** (5 cm)

Paine 1998), temperature (Tomanek and Helmuth 2002; Helmuth et al. 2006; Lathlean et al. 2011), salinity (Zacharias and Roff 2001), substrate orientation (Chapman and Bulleri 2003), topography heterogeneity (Archambault and Bourget 1996; Cusson and Bourget 1997; Guichard and Bourget 1998; Guichard et al. 2001), predation and competition (Connell 1961a), disturbance and succession (Sousa 1979a, b), grazing (Apolinário et al. 1999; Bulleri et al. 2000; Sala et al. 2011), and recruitment (Forde and Raimondi 2004; Skinner and Coutinho 2005). Their relative importance is hard to generalize, due to the frequently high number of potential factors involved and, mainly, their interactions (Little and Kitching 1996).

Benthic community zonation patterns are well known, and this subject has been discussed by several authors (Underwood 1978; Coutinho 1995; Benedetti-Cecchi and Cinelli 1997; Good 2004; Coutinho and Zalmon 2009). For marine biologists, the zones are extremely useful in a practical sense because they provide a convenient descriptive framework into which specific observations can be fitted (Little and Kitching 1996).

On rocky shores, substrate topography may have a spatial influence on the intensity of environmental variables, which, in turn, might affect the biological processes during both low and high tide periods (Guichard et al. 2001). The physical features of the environment and consequently the structure of the local assemblage may change abruptly over a very small spatial scale in high complexity habitats (Benedetti-Cecchi and Cinelli 1997; Benedetti-Cecchi et al. 2003).

Increasing environmental stress has historically been related to lower diversity, richness, and higher species dominance (Clarke and Warwick 2001), since the physical factors such as heat and desiccation are more severe due to the longer exposure time. However, tide oscillations at the intermediate levels of the substrate turned the environment into a less stressful one and allow several organisms to inhabit this zone. Thus, biological factors such as species interactions (i.e. competition for limited resources mainly space, predation and grazing; Fig. 4.1g, h) are also important determinants of zonation.

Therefore, once we already know the environment and organisms, this chapter intends to deepen into the post-recruitment interaction processes between benthic species (competition, predation, and herbivory) and how this leads to zonation and community dynamics. The question here is: What are the main effects of biotic (i.e. competition, predation, herbivory, or facilitation) and abiotic drivers (i.e. water temperature, air temperature, tidal regime, water quality, and others) in the distribution pattern of rocky intertidal species?

Post-settlement Interactions

Subtidal communities are often viewed as extensions of their intertidal counterparts (Witman and Dayton 2001). Important processes within intertidal systems such as (1) periodic disturbances of the community (e.g. Dayton 1971; Sousa 1979a, b, 1980), (2) strong competitive and trophic relationships controlled by one or a few key species (e.g. Connell 1961a, 1961b; Paine 1966; Dayton 1971; Underwood 1981) and (3) recruitment into open space which is variable and often a function of large-scale hydrographic phenomena (e.g. Shanks 1983; Wethey 1985; Roughgarden et al. 1988; Shkedy and Roughgarden 1997) are also seen as important in the subtidal communities (e.g. Sutherland 1974; Osman 1977, 1978; Moran 1980; Ayling 1981; Witman 1985). Because many of the species in these systems are relatively short-lived, recruitment is a critical component. Competition and predation may control dominance, and disturbance may establish the extent of free space, but recruitment controls which species are present and their initial abundances.

According to Osman and Whitlatch (1995), post-settlement predation needs to be recognized as a generally important process affecting community composition. For those systems in which recruitment is a dominant or controlling process, post-settlement predation may often be the most important process. In this light, the intertidal barnacle–mussel communities, in which supply often seems to control recruitment (e.g. Gaines and Roughgarden 1985; Raimondi 1988; Sutherland 1990; Minchinton and Scheibling 1991), may be viewed as an end member of a continuum. In such harsh environments, the diversity of possible prey and predators is greatly reduced, limiting both the potential number of post-settlement predators and their ability to control recruitment. However, even within intertidal systems, there have been indications that predation on recruiting life-stages can be important (e.g. Palmer 1990; Gosselin and Qian 1996).

Post-recruitment Interactions: Abiotic and Biotic

Both abiotic and biotic processes influence rocky community structure (Menge and Sutherland 1987), and the interactions among them are responsible for the non-homogeneous patterns of population distribution in the intertidal zone (Murray et al. 2006; Zamprogno et al. 2012). All these different processes can affect the species distribution at different temporal scales (Benedetti-Cecchi 2000).

The physical structure of the habitat can alter susceptibility of intertidal animals to desiccation stress, wave exposure and predation by creating potential "niches" (Sebens 1986). On tropical rocky shores, temperature and desiccation stress are important factors structuring the intertidal community (Garrity 1984; Williams 1993). Many mobile species reduce thermal stress by hiding in rock crevices and cracks, which provide shadow and refuges (Garrity 1984; Williams and Morritt 1995). Aggregation, selection of habitat, decreasing shell pigmentation and increased/decreased shell ornamentation are some adaptations of *Littorina* to survive in this harsh environment where the desiccation stress is the highest (Chapman and Underwood 1996). The densities of many rocky shore gastropods are, thus, affected positively by higher habitat complexity, but the opposite is shown for *Littorina saxatilis* (Olivi, 1792). Smaller morphs of this littorinid snail occur in higher densities in the less complex barnacle habitat than the more complex mussel habitat, which is likely to be related to small refuges in between barnacle shells

(Kostylev 1996; Kostylev et al. 1997). Also, Masi and Zalmon (2008) in breakwaters at the north Rio de Janeiro showed higher abundance of the bivalvia *Perna perna* (Linnaeus, 1758) on exposed rocky shores as result of lower predation rate by the carnivorous gastropod *Stramonita haemastoma* (Linnaeus, 1767), a common species hidden at crevices on shelterer rocky places.

Puga et al. (2019) showed an interesting result in Guanabara Bay, RJ, with a significant negative relationship (competition) between the non-indigenous oyster species *Saccostrea cucculata* (Born, 1778) and the historically introduced barnacle species *Amphibalanus amphitrite* (Darwin, 1854). It seems that while this species seems to be quite resistant to several environmental conditions that are unfavorable to many species, when the environmental conditions become favorable to some of these species, *A. amphitrite* did not get to compete and is restricted by the preemption of substratum by other species. A similar pattern is observed by these authors in the Paquetá rocky shore in this same bay, where the pollution is one of the unfavorable environmental conditions that limits the distribution of mid-intertidal species. In this area, it is possible to see higher percent cover of *A. amphitrite* in some seasons in contrast to lower oyster percent cover, which seems to be related to its resistance and its weak ability as competitor.

Populations and communities can also change because of emigration and immigration—usually referred to as dispersal and recruitment or settlement, at least for species with propagules. Also, the facilitation observed among benthic organisms is well established in the literature, since most of them are ecosystem engineers, which means that they can increase habitat complexity and heterogeneity, important factors in structuring rocky shore communities (Underwood et al. 1983; Menge et al. 1986a). Studies with the oyster *Crassostrea gigas* (Thunberg, 1793) verified that the presence of "neighbors" surrounding the oyster, regardless the species identity, improved the survival ameliorating the effect of harsher environmental conditions (Ruesink 2007; Hedge and Johnston 2014).

Higher flow rates, often verified at exposed sites, are responsible for increased delivery of food resources and larval supply (both considered bottom-up drivers), allowing a higher abundance of filter-feeding organisms (e.g. Leonard et al. 1998). Because filter-feeding invertebrates are dominant competitors by space (e.g. Menge et al. 1986b; Bell 2008), they decrease the abundance of seaweeds in exposed shores, as observed in Ubatuba by Christofoletti et al. (2011). According to these authors, the abundance of sessile organisms varied according to wave exposure gradient. Sheltered shores were dominated by primary producers, while filter-feeding animals prevailed in more exposed areas, probably because they overcompete with seaweeds; the high abundance of sessile invertebrates in exposed areas seemed to be the main factor increasing the abundance of predators at the sublittoral fringe. In summary, the spatial distribution of sessile organisms and sedentary consumers in the study area suggests both physical and biological control of community dynamics.

The mosaic structure of the intertidal community is primarily a consequence of disturbance, competition for space among neighboring mosaic elements, life history differences, colonization ability, and how these vary with wave exposure. Differences in species composition with wave-exposure regime undoubtedly reflect differences in physiological tolerances and for some species, probably dispersal limitation. Life history differences, including morphological characteristics such as whether or not a species has an encrusting, spreading holdfast system, also seem to play a role in determining whether or not an alga is a common component of the algal mosaic.

In this environment, top-down control by sedentary consumers plays an important role on populations of prey, since the grazing activity of limpets and sea urchins influence the abundance and distribution of microalgae and seaweeds (e.g. Andrew 1993; Benedetti-Cecchi et al. 2001; Jenkins et al. 2001, 2005; Coleman et al. 2006; Scheibling et al. 2009), while whelks and sea stars negatively impact populations of sessile invertebrates (e.g. Paine 1971; Fairweather and Underwood 1991; Navarrete 1996; Navarrete and Menge 1996). In addition to top-down control, wave exposure also influences abundance and distribution of prey, both seaweeds and filter-feeding animals. Wave action directly affects the abundance of filter-feeding animals due to the positive influence of flow rate on feeding and settlement rates and mitigation of desiccation stress (e.g. Leonard et al. 1998, 1999). Also, waves indirectly affect populations of prey by controlling the distribution of sedentary consumers. Higher hydrodynamic action decreases emersion stress but increases dislodgement risk of animals due to wave impact, consequently, density and aggregation of consumers are increased near crevices where foraging pressure is particularly high (e.g. Kensler 1967; Johnson et al. 1998; Rilov et al. 2005; Stafford and Davies 2005; Jonsson et al. 2006; Stafford et al. 2007). Therefore, diversity, distribution, competition, and abundance of both seaweeds and animals on intertidal rocky shores are driven by the interplay between physical factors and consumer stresses.

According to Connell and Slatyer (1977), the influence of consumers is also of critical importance to the course of succession in marine intertidal communities, and several studies indicate that consumers often have a strong influence on the rate of succession, accelerating (Lubchenco and Menge 1978; Sousa 1979a, b, 1984; Day and Osman 1981), unaffecting (Turner 1983; Jemakoff 1985) or slowing (Dayton 1975; Sousa et al. 1981; Farrell 1991). In an upper rocky intertidal shore of British Columbia, Canada, Kim and DeWreede (1996) indicated that algal succession followed an inhibition model, in which the early colonist, ephemeral seaweed, inhibited the settlement of the later species, *Fucus gardneri* P.C. Silva and *Pelvetiopsis limitata* (Setchell) N.L.Gardner, for the first 2 years. However, this inhibition occurred only in the absence of limpets. In the limpet-present plots, limpets grazed a substantial amount of ephemeral seaweed, and consequently fucoids occupied space in the early stage of succession, and this seaweed persisted. The positive effect of limpets on fucoids is an indirect facilitation mediated by ephemeral seaweed, which may occur because the limpets prefer to eat ephemerals rather than fucoids. The indirect effect of limpets on fucoid seaweed was a key interaction in understanding community dynamics in this habitat.

In summary, because many of the species in intertidal rocky shore systems are relatively short-lived, recruitment is a critical component. Competition and predation may control dominance, and disturbance may establish the extent of free space, but recruitment controls which species are present and their initial abundances. For those systems in which recruitment is a dominant or controlling process, post-settlement predation may often be the most important process. However, in such harsh environments, the diversity of possible prey and predators is greatly reduced, limiting both the potential number of post-settlement predators and their ability to control recruitment, although within intertidal systems there have been indications that predation on recruiting life-stages can be important. Also, the interactions among biotic and abiotic processes are responsible for the non-homogeneous patterns of population distribution in the intertidal zone and can affect the species distribution at different temporal scales. Populations and communities can also change because of emigration and immigration, and the facilitation observed among benthic organisms is well known, since most of them are ecosystem engineers, which means that they can increase habitat complexity and heterogeneity, important factors in structuring rocky shore communities. Finally, the mosaic structure of the intertidal community is primarily a consequence of disturbance, competition for space among neighboring mosaic elements, life history differences, colonization ability, and how these vary with wave exposure. Differences in species composition with wave-exposure regime undoubtedly reflect differences in physiological tolerances and for some species, probably dispersal limitation. In this environment, top-down control by sedentary consumers plays an important role on populations of prey, since the grazing activity of limpets and sea urchins influence the abundance and distribution of microalgae and seaweeds. Therefore, diversity, distribution, competition and abundance of both seaweeds and animals on intertidal rocky shores are driven by the interplay between physical factors and consumer stresses.

References

Andrew NL (1993) Spatial heterogeneity, sea-urchin grazing, and habitat structure on reefs in temperate Australia. Ecology 74:292–302

Apolinário M, Coutinho R, Baeta-Neves MH (1999) Periwinkle (Gastropoda, Littorinidae) habitat selection and its impact upon microalgal populations. Rev Bras Biol 59:211–219

Archambault P, Bourget E (1996) Scales of coastal heterogeneity and benthic intertidal species richness, diversity and abundance. Mar Ecol Prog Ser 36:111–121

Ayling AM (1981) The role of biological disturbance in temperate subtidal encrusting communities. Ecology 62:830–847

Bell JJ (2008) The functional roles of marine sponges. Estuar Coast Shelf Sci 79:341–353

Benedetti-Cecchi L (2000) Predicting direct and indirect effects during succession in a midlittoral rocky shore assemblage. Ecol Monogr 70:45–72

Benedetti-Cecchi L, Cinelli F (1997) Spatial distribution of algae and invertebrates in the rocky intertidal zone of the Strait of Magellan: are patterns general. Polar Biol 18:337–343

Benedetti-Cecchi L, Bulleri F, Acunto S, Cinelli F (2001) Scales of variation in the effects of limpets on rocky shores in the northwest Mediterranean. Mar Ecol Prog Ser 209:131–141

Benedetti-Cecchi L, Maggi E, Bertocci I, Vaselli S, Micheli F, Osioa GC, Cinelli F (2003) Variation in rocky shore assemblages in the northwestern Mediterranean: contrasts between islands and the mainland. J Exp Mar Biol Ecol 293:193–215

Bulleri F, Menconi M, Cinelli F, Benedetti-Cecchi L (2000) Grazing by two species of limpets on artificial reefs in the northwest Mediterranean. J Exp Mar Biol Ecol 255:1–19

Chapman MG, Bulleri F (2003) Intertidal seawalls: new features of landscape in intertidal environments. Landsc Urban Plan 62:159–172

Chapman MG, Underwood AJ (1996) Influences of tidal conditions, temperature and desiccation on patterns of aggregation of the highshore periwinkle, *Littorina unifasciata*, in New South Wales, Australia. J Exp Mar Biol Ecol 196:213–237

Christofoletti RA, Takahashi CK, Oliveira DN, Flores AAV (2011) Abundance of sedentary consumers and sessile organisms along the wave exposure gradient of subtropical rocky shores of the south-west Atlantic. J Mar Biol Assoc UK 91:961–967

Clarke KR, Warwick RM (2001) Change in marine communities: an approach to statistical analysis and interpretation, 2nd edn. PRIMER-E, Plymouth

Coleman RA, Underwood AJ, Benedetti-Cecchi L, Aberg P, Arenas F, Arrontes J, Castro J, Hartnoll RG, Jenkins SR, Paula J, Della Santina P, Hawkins SJ (2006) A continental scale evaluation of the role of limpet grazing on rocky shores. Oecologia 147:556–564

Connell JH (1961a) The influence of interspecific competition and other factors on the distribution of the barnacle *Chthamalus stellatus*. Ecology 42:710–723

Connell JH (1961b) Effects of competition, predation by *Thais lapillus*, and other factors on natural populations of the barnacle *Balanus balanoides*. Ecol Monogr 31:61–104

Connell JH, Slatyer RO (1977) Mechanisms of succession in natural communities and their role in community stability and organization. Am Nat 111:1119–1144

Coutinho R (1995) Avaliação crítica das causas da zonação dos organismos bentônicos em costões rochosos. Oecol Brasil 1:259–271

Coutinho R, Zalmon IR (2009) O bentos de costões rochosos. In: Pereira RC, Soares-Gomes A (eds) Biologia Marinha. Cap. 11. Interciência, Rio de Janeiro, pp 281–297

Cusson M, Bourget E (1997) Influence of topographic heterogeneity and spatial scales on the structure of the neighbouring intertidal endobenthic macrofaunal community. Mar Ecol Prog Ser 150:181–193

Day RW, Osman RW (1981) Predation by *Petiria miniata* (Asteroidea) on bryozoans: prey diversity may depend on the mechanism of succession. Oecologia 51:300–309

Dayton PK (1971) Competition, disturbance, and community organization: the provision and subsequent utilization of space in a rocky intertidal community. Ecol Monogr 41:351–388

Dayton PK (1975) Experimental evaluation of ecological dominance in a rocky intertidal algal community. Ecol Monogr 45:137–159

Denny MW, Paine RT (1998) Celestial mechanics, sea-level changes, and intertidal ecology. Biol Bull 194:108–115

Fairweather PG, Underwood AJ (1991) Experimental removals of a rocky intertidal predator: variations within two habitats in the effects on prey. J Exp Mar Biol Ecol 154:29–75

Farrell TM (1991) Models and mechanisms of succession: an examplefrom a rocky intertidal community. Ecol Monogr 61:95–113

Forde S, Raimondi PT (2004) An experimental test of the effects of variation in recruitment intensity on intertidal community composition. J Exp Mar Biol Ecol 301:1–14

Gaines S, Roughgarden J (1985) Larval settlement rate: a leading determinant of structure in an ecological community of the marine intertidal zone. Proc Natl Acad Sci USA 82:3707–3711

Garrity SD (1984) Some adaptations of gastropods to physical stress on a tropical rocky shore. Ecology 65:559–574

Good TP (2004) Distribution and abundance patternsin caribbean rocky intertidal zones. Bull Mar Sci 74:459–468

Gosselin LA, Qian PY (1996) Early post-settlement mortality of an intertidal barnacle: a critical period for survival. Mar Ecol Prog Ser 135:69–75

Guichard F, Bourget E (1998) Topographic heterogeneity, hydrodynamics, and benthic community structure: a scale-dependent cascade. Mar Ecol Prog Ser 171:59–70

Guichard F, Bourget E, Robert JL (2001) Scaling the influence of topographic heterogeneity on a intertidal benthic communities: alternate trajectories mediated by hidrodinamics and shading. Mar Ecol Prog Ser 217:27–41

Harley GDC, Anderson MK, Demes WK, Jorve PJ, Kordas LR, Coyle AT (2012) Effects of climate change on global seaweed communities. J Phycol 48:1064–1078

Hedge LH, Johnston EL (2014) Colonisation of the non-indigenous pacific oyster *Crassostrea gigas* determined by predation, size and initial settlement densities. PLoS One 9(3):e90621

Helmuth B, Mieszkowska N, Moore P, Hawskins SJ (2006) Living on the edge of two changing worlds: forecasting the responses of rocky intertidal ecosystems to climate change. Annu Rev Ecol Evol Syst 37:423–431

Hershner C, Mason P (2009) Ecosystems services of tidal shorelines. Rivers Coast 4:1–8

Horn MH, Martin KLM, Chotkowski MA (1999) Intertidal fishes – introduction. Academic Press, San Diego

Jemakoff P (1985) The effect of overgrowth by algae on the survival of the intertidal barnacle *Tesseropora rosea* Krauss. J Exp Mar Biol Ecol 94:89–97

Jenkins SR, Arenas F, Arrontes J, Bussell J, Castro J, Coleman RA, Hawkins SJ, Kay S, Martinez B, Oliveros J, Roberts MF, Sousa S, Thompson RC, Hartnoll RG (2001) European-scale analysis of seasonal variability in limpet grazing activity and microalgal abundance. Mar Ecol Prog Ser 211:193–203

Jenkins SR, Coleman RA, Della Santina P, Hawkins SJ, Burrows MT, Hartnoll RG (2005) Regional scale differences in the determinism of grazing effects in the rocky intertidal. Mar Ecol Prog Ser 287:77–86

Johnson MP, Burrows MT, Hawkins SJ (1998) Individual based simulations of the direct and indirect effects of limpets on a rocky shore *Fucus* mosaic. Mar Ecol Prog Ser 169:179–188

Jonsson PR, Granhag L, Moschella PS, Aberg P, Hawkins SJ, Thompson RC (2006) Interactions between wave action and grazing control the distribution of intertidal macroalgae. Ecology 87:1169–1178

Kensler CB (1967) Desiccation resistance of intertidal crevice species as afactor in their zonation. J Anim Ecol 36:391–406

Kim JH, DeWreede RE (1996) Effects of size and season of disturbance on algal patch recovery in a rocky intertidal community. Mar Ecol Prog Ser 133:217–228

Kostylev V (1996) Spatial heterogeneity and habitat complexity affecting marine littoral fauna. Ph.D. Thesis, Göteborg University, Sweden

Kostylev V, Erlandsson J, Johannesson K (1997) Micro distribution of the polymorphic snail *Littorina saxatilis* (Olivi) in a patchy rocky shore habitat. Ophelia 47:1–12

Lathlean JA, Ayre DJ, Todd E (2011) Minchinton rocky intertidal temperature variability along the southeast coast of Australia: comparing data from *in situ* loggers, satellite-derived SST and terrestrial weather stations. Mar Ecol Prog Ser 439:83–95

Leonard GH, Levine JM, Schmidt PR, Bertness MD (1998) Flow-driven variation in intertidal community structure in a Maine estuary. Ecology 79:1395–1411

Leonard GH, Ewanchuk PJ, Bertness MD (1999) How recruitment, intraspecific interactions, and predation control species borders in a tidal estuary. Oecologia 118:492–502

Little C, Kitching JA (1996) The biology of rocky shores. Oxford University Press, New York

Lubchenco J, Menge BA (1978) Community development and persistence in a low rocky intertidal zone. Ecol Monogr 48:67–94

Masi BP, Zalmon IR (2008) Zonação da comunidade bêntica do entremarés em molhes sob diferente hidrodinamismo na costa norte do estado do Rio de Janeiro, Brasil. Rev Bras Zool 25:662–673

Menge BA, Sutherland JP (1987) Community regulation: variation in disturbance, competition, and predation in relation to environmental stress and recruitment. Am Nat Mag 130:730–757

Menge BA, Lubchenco J, Gaines SD (1986a) A test of the Menge-Sutherland model of community organization in a tropical rocky intertidal food web. Oecologia 71:75–89

Menge BA, Lubchenco J, Ashkenas LR, Ramsey F (1986b) Experimental separation of effects of consumers on sessile prey inthe low zone of a rocky shore in the Bay of Panama—direct and indirect consequences of food web complexity. J Exp Mar Biol Ecol 100:225–269

Minchinton TE, Scheibling RE (1991) The influence of larval supply and settlement on the population structure of barnacles. Ecology 72:1867–1879
Moran J (1980) Natural physical disturbance and predation: their importance in structuring a marine sessile community. Aust J Ecol 5:193–200
Murray SN, Ambrose RF, Dethier MN (2006) Monitoring rocky shores. University of California Press, California
Navarrete SA (1996) Variable predation: effects of whelks on a midintertidal successional community. Ecol Monogr 66:301–321
Navarrete AS, Menge BA (1996) Keystone predation and interaction strength: interactive effects of predators on their main prey. Ecol Monogr 66:409–429
Osman RW (1977) The establishment and development of a marine epifaunal community. Ecol Monogr 47:37–63
Osman RW (1978) The influence of seasonality and stability on the species equilibrium. Ecology 59:383–399
Osman RW, Whitlatch RB (1995) Predation on early ontogenetic life-stages and its effect on recruitment into a marine community. Mar Ecol Prog Ser 117:111–126
Paine RT (1966) Food web complexity and species diversity. Am Nat 100:65–75
Paine RT (1971) Short-term experimental investigation of resource partitioning in a New Zealand rocky intertidal habitat. Ecology 52:1096–1106
Palmer AR (1990) Predator size, prey size, and the scaling of vulnerability: hatchling gastropods vs. barnacles. Ecology 71:759–775
Puga C, Torres ASS, Paiva PC, Yoneshigue-Valentin Y, Junqueira AOR (2019) Multi-year changes of a benthic community in the mid-intertidal rocky shore of a eutrophic tropical bay (Guanabara bay, RJ – Brazil). Estuar Coast Shelf Sci 226:106265
Raimondi PT (1988) Rocky type affects settlement, recruitment, and zonation of the barnacle *Chthamalus anisopoma* Pilsbury. J Exp Mar Biol Ecol 123:253–267
Rilov G, Gasith A, Benayahu Y (2005) Effect of disturbance onforaging: whelk activity on wave-exposed rocky shores with minimal tidal range. Mar Biol 147:421–428
Roughgarden J, Gaines S, Possingham H (1988) Recruitment dynamics in complex life cycles. Science 241:1460–1466
Ruesink JL (2007) Biotic resistance and facilitation of a nonnative oyster on rocky shores. Mar Ecol Prog Ser 331:1–9
Sala E, Kizilkaya Z, Yildirim D, Ballesteros E (2011) Alien marine fishes deplete algal biomass in the Eastern Mediterranean. PLoS One 6(2):e17356. https://doi.org/10.1371/journal.pone.0017356
Scheibling RE, Kelly NE, Raymond BG (2009) Herbivory and community organization on a subtidal cobble bed. Mar Ecol Prog Ser 382:113–128
Sebens K (1986) Spatial relationships among encrusting marine organisms in the New England subtidal zone. Ecol Monogr 56:73–96
Shanks AL (1983) Surface slicks associated with tidally forced internal waves may transport pelagic larvae of benthic invertebrates and fishes shoreward. Mar Ecol Prog Ser 13:311–315
Shkedy Y, Roughgarden J (1997) Barnacle recruitment and population dynamics predicted from coastal upwelling. Oikos 80:487–498
Skinner LF, Coutinho R (2005) Effect of microhabitat distribution and substrate roughness on barnacle *Tetaraclita stalactifera* (Lamarck, 1818) settlement. Braz Arch Biol Technol 48:109–113
Sousa WP (1979a) Experimental investigations of disturbance and ecological succession in a rocky intertidal algal community. Ecol Monogr 49:227–254
Sousa WP (1979b) Disturbance in marine intertidal boulders fields: the non equilibrium maintenance of species diversity. Ecology 60:1225–1239
Sousa WP (1980) The responses of a community to disturbance: the importance of successional age and species' life histories. Oecologia 45:72–81
Sousa WP (1984) Intertidal mosaics: patch size, propagule availability, and spatially variable patterns of succession. Ecology 65:1918–1935

Sousa WP, Schroeter SC, Gaines SD (1981) Latitudinal variation in intertidal algal community structure: the influence of grazing vegetative propagation. Oecologia 48:297–307
Stafford R, Davies MS (2005) Spatial patchiness of epilithic biofilm caused by refuge-inhabiting high shore gastropods. Hydrobiologia 545:279–287
Stafford R, Davies MS, Williams GA (2007) Computer simulations of high shore littorinids predict small-scale spatial and temporal distribution patterns on rocky shores. Mar Ecol Prog Ser 342:151–161
Sutherland JP (1974) Multiple stable points in natural communities. Am Nat 108:859–873
Sutherland JP (1990) Recruitment regulates demographic variation in a tropical intertidal barnacle. Ecology 71:955–972
Tomanek L, Helmuth B (2002) Physiological ecology of rocky intertidal organisms: a synergy of concepts. Integr Comp Biol 42:771–775
Turner T (1983) Facilitation as a successional mechanism in a rocky intertidal community. Am Nat 121:729–738
Underwood AJ (1978) A refutation of critical tidal levels as determinants of the structure of intertidal communities on British shores. J Exp Mar Biol Ecol 33:261–276
Underwood AJ (1981) Structure of a rocky intertidal community in New South Wales: patterns of vertical distribution and seasonal changes. J Exp Mar Biol Ecol 37:857–870
Underwood AJ, Denley EJ, Moran MJ (1983) Experimental analyses of the structure and dynamics of mid-shore rocky intertidal communities in New South Wales. Oecologia 56:202–219
Wethey DS (1985) Catastrophe, extinction, and species diversity: a rocky intertidal example. Ecology 66:445–456
Williams GA (1993) Seasonal variation in algal species richness and abundance in the presence of molluscan herbivores on a tropical rocky shore. J Exp Mar Biol Ecol 167:261–275
Williams GA, Morritt D (1995) Habitat partitioning and ther-mal tolerance in a tropical limpet, Cellana grata. Mar Ecol Prog Ser 124:89–103
Witman JD (1985) Refuges, biological disturbance, and rocky subtidal community structure in New England. Ecol Monogr 55:421–445
Witman JD, Dayton PK (2001) Rocky subtidal communities. In: Bertness MD, Gaines SD, Hay ME (eds) Marine community ecology. Sinauer Associates, Sunderland, pp 339–366
Zacharias MA, Roff JC (2001) Explanations of patterns of intertidal diversity at regional scales. Biogeographica 28:471–483
Zamprogno GC, Fernandes FC, Fernandes LL (2012) Iheringia. Fortschr Zool 102:375–383

Chapter 5
Economic and Cultural Aspects of Ecosystem Services in Rocky Shores

Natalia Pirani Ghilardi-Lopes, Mariana Clauzet, Juliana Imenis Barradas, and Valéria Vinha

Abstract Ecosystem services offered by rocky shores significantly enhance human well-being. However, these benefits are increasingly at risk due to the rising demand for recreational resources and tourism facilities in coastal areas, compounded by the threats of sea level rise and coastal erosion linked to climate change. These coastal ecosystems are crucial for providing food and habitat for fish, and they sustain artisanal fishing communities that utilize low-draft vessels and traditional fishing gear. These communities have developed a profound understanding of local fisheries, enabling them to live in harmony with their environment. However, they are increasingly exposed to the pressures threatening coastal resources. In response, they have formed alliances to secure legal recognition for what have often been overlooked as marginal economies, advocating for defined property rights and resource access. Additionally, rocky shores are celebrated for their rich marine biodiversity and breathtaking landscapes, offering extensive recreational and educational opportunities. These areas function as natural classrooms and therapeutic retreats, fostering conservation and improving community well-being. They also serve as centers of cultural heritage, inspiring artists and preserving traditional practices within local communities. This chapter delves into the complex role of rocky shores in fostering economic and cultural ecosystem services.

N. P. Ghilardi-Lopes (✉)
Center for Natural and Human Sciences, Federal University of ABC, Postgraduate Program in Teaching and History of Science and Mathematics (PEHFCM), São Paulo, Brazil
e-mail: natalia.lopes@ufabc.edu.br

M. Clauzet
National Institute of Science and Technology in Public Policies, Strategies and Development (INCT-PPED), Rio de Janeiro, Brazil

J. I. Barradas
Center for Natural and Human Sciences, Federal University of ABC, Postgraduate Program in Teaching and History of Science and Mathematics (PEHFCM), São Paulo, Brazil

Municipal University of São Caetano do Sul, São Paulo, Brazil

V. Vinha
Graduate Program in Public Policies, Strategies and Development at Federal University of Rio de Janeiro, Institute of Economics (PPED/IE/UFRJ), Rio de Janeiro, Brazil

R. Coutinho, R. A. Christofoletti (eds.), *Brazilian Rocky Shores*, Brazilian Marine Biodiversity, https://doi.org/10.1007/978-3-031-67206-4_5

Valuation and Use of Ecosystem Services Provided by Rocky Shorelines: Implications for the Conservation of Coastal Socioecosystems

The ecosystem services and benefits to human well-being provided by coastal environments and by rocky shores or intertidal rock formations are associated with provision of biological, atmospheric, and climatic regulation; flood and storm protection; ecosystem maintenance; nutrient cycling; recreation and tourism; scenic quality, sense of place and spiritual values, education, and research (UNEP 2006; Elliff and Kikuchi 2015; Garcia 2019).

Rocky shores provide goods and services to humans, such as shellfish and seaweeds that have furnished subsistence foods since prehistory, as well as natural coastal defense and recreational opportunities. They are important feeding areas for birds and fish. Many of the species in these habitats are endemic (notably on Australia's southern coast—Clark et al. 2021). As a productive food source, rocky shorelines provide an important nursery area for many fish and crustacean species. Some species seek refuge along rocky shores, where dense stands of seaweed help to diminish the force of the waves. These algal beds provide a crucial habitat and serve as a vital food source for rare and threatened species, including marine turtles (Clark et al. 2021).

Beyond providing essential food security for coastal fishers and shellfish gatherers, rocky shores offer a multitude of services. They serve as habitats and anchorage points, support foundations for human constructions, and provide shelters for ancient settlements. Additionally, these areas function as natural basins for the control and storage of water. The environmental resilience of intertidal rocks is enhanced by their natural ability to protect against erosion, effectively reducing the wave energy that reaches the shore. Ecologically, communities on intertidal rocks demonstrate a strong capacity to naturally recover from anthropogenic impacts. This resilience is largely due to the influx of propagules from unaffected areas, which help replenish and restore their ecological balance (Crowe et al. 2000). Intertidal rock habitats are important sources of larval plankton upon which commercially important fish species feed, including mussels and larval fish of plaice and mackerel. Additionally, these habitats support the commercial harvesting of various rocky shore species that humans exploit, including kelp, seaweed, edible crabs, mussels, and winkles (Crowe et al. 2000).

Importance to Subsistence Fishers and Conflicts with Competing Uses

Artisanal fishing activities in coastal zones, along with resource collection in mangroves, salt flats, and rocky shores, engage over one million workers directly and an additional three million indirectly around the globe (Mattos et al. 2020), ensuring food security and the maintenance of an invaluable cultural and symbolic heritage (Berkes et al. 2000; Johannes et al. 2000).

The collection of resources (shellfish, crabs, oysters, etc.) occurs predominantly in areas of rocky shores and mangroves, through manual collection and/or the use of traditional equipment. In many regions, the activity is mainly carried out by women, who play a vital role in the social movements fighting for fishing rights. According to the recommendations on gender equality of the FAO Guidelines for Small-Scale Fisheries, this reality needs to be made visible and reinforced through public gender policies that guarantee the role of women in the development of small-scale fisheries activities (FAO 2015).

Although dependent on the environmental integrity of ecosystems, which constitutes a difficulty in the face of climate impacts on the coastal-marine environment, artisanal fishing and gathering compete with other potentially impacting economic activities in the coastal-marine zone such as port facilities, oil and gas industries, real estate speculation, disorderly nautical tourism, etc. (Mendonça et al. 2018). Therefore, it is extremely important to investigate and recognize the socioeconomic dependence and cultural heritage associated with these activities to insulate the various uses in this territory from those which disrupt them.

Rocky shores located in the most populated and industrialized regions are often the focus of extreme real-estate speculation and a tourism industry that lacks environmental safeguards despite the existence of legislation requiring participatory impact assessment and mitigation, thus endangering both the quantity and the quality of coastal ecosystem services (Garcia et al. 2018). Rocky shorelines near population centers are threatened by trampling, harvesting, bait collecting, artificial light pollution, and poor water quality. They are one of the habitats most affected by urbanization, which causes loss and fragmentation. Regulated closures, such as marine parks, offer some protection, but poaching threatens their effectiveness. Designating rocky shores as protected areas can sometimes lead to conflicts with artisanal fishers who rely on access to these areas for their livelihoods. Despite their minimal impact on these environments, the restriction of access can significantly affect their traditional fishing practices and economic stability (see works by Diegues; Begossi and others). Another impact is related to pollutants, which can enter rocky shores as waterborne (e.g., petroleum, microplastics) or land-based contaminants, potentially causing toxicity to some organisms (Mearns et al. 2020).

Current rates of rapid demographic change—an increasing and aging population, who have more leisure time—are expected to bring about an even more increased demand for coastal recreation. This will mean more demand for land for housing and recreational facilities and may also be associated with conflicting development of ports and renewable energy facilities such as windfarms (Hadley 2009).

Artisanal fishing is multispecific (diverse in catch) and relies mainly on family labor, and its production is associated with subsistence and/or the commercial market and can be a seasonal alternative for practitioners who dedicate themselves part of the year to agriculture (fisher/farmer) or to other economic activities such as tourism (Diegues 1983, 1995; Bayley and Petrere Jr. 1989).

Complementarity Between Traditional Resource Management and Ecosystem Services

These social groups reinforce the importance of recognition of ecosystem services provision to the benefit of human societies. These benefits are derived from cultural features associated with traditional fishing and shellfish gathering, including:

(i) The management of natural resources is based on the application of local ecological knowledge (or traditional knowledge), transmitted orally and elaborated by local innovations through observation/experimentation over generations.
(ii) Use of vessels and capture and collection equipment with low navigation/displacement autonomy and capture capacity.
(iii) Purpose of the activity directly related to food security.

In this sense, such activities, far from impacting biodiversity, in most cases maintain favorable or functional environmental conditions for the equilibrium or resilience of coastal-marine biodiversity. Likewise, the implementation of community-based tourism services in coastal communities through properly organized participatory processes also contributes to the functioning and maintenance of coastal-marine ecosystems.

However, artisanal fishermen and gatherers are extremely vulnerable to environmental impacts, as the exploited environments are limited to the proximity between the coast and the fisherman's home and the islands located in the bays close to the coastal zone. Neither collectors nor artisanal fishermen have the autonomous capacity to relocate or reorient their activities to areas farther from the coastal zone in case of natural disasters and environmental impacts; their survival is strongly linked with ancestral knowledge and traditional property rights over local fishing points recognized in community management systems (Acheson 1981; Begossi 1995, 2006; Clauzet et al. 2007; Clauzet 2009; Begossi et al. 2013; Lopes et al. 2015). In this context, it is imperative that the use of the coastal-marine territory by dependent social groups be strengthened through use and harvest rights and upheld by appropriate legislation and local covenants.

Threats Due to Climate Change

The most significant threats to the health of rocky shoreline ecosystems primarily arise as consequences of climate change. These include changes in ocean chemistry and temperature, resulting in changes in the composition and distribution of species inhabiting rocky shores or leading to significant species mortality. Climate-related storm activity is likely to mobilize sediments, which may abrade or bury some species (Coutinho et al. 2016). Finally, sea level rise poses a significant threat to rocky shores by reducing suitable habitat; as water levels increase, many of the platforms that constitute these shores become inundated, diminishing the available space for species to thrive (Starko et al. 2019; Mieszkowska et al. 2021). The proximate effects of sea level rise, driven by the melting of Arctic and Greenland ice packs, are considerable for the health of rocky coastal habitats.

Both UNEP, through the Millennium Ecosystem Assessment, and the IPCC special report on coastal ecosystem impacts have dedicated specific analyses to the potential dangers that would ensue from predicted warming scenarios. From a global perspective, the comprehensive IPCC report on the impact of climate change on coastal ecosystems describes the most salient indicators (IPCC 2022). The impacts of global warming on fisheries and their important role in food security are observed in declining fish stocks which are also depleted by overfishing. Indigenous fishing communities that depend on traditional marine resources for food and economic security are particularly vulnerable to climate change because of reduced capacity to carry out traditional harvests owing to reduced access to, or availability of, resources. This could occur due to conversion of rocky shores for other uses or as a result of sea level rise.[1]

Legislation to Strengthen Artisanal Fishing Rights and Access

To avoid usurpation of the rights exercised by traditional fishers and coastal producers, legislation and recognition of traditional use rights are essential. In Brazil, legislation is under consideration (Bill PL-6969/2013[2]), which would establish a National Policy for the Conservation and Sustainable Use of the Brazilian Marine Biome (PNCMar), while a companion proposal (PL 131/2020[3]) provides for the recognition and protection and guarantees to territorial rights of traditional fishing communities. Such rights are to be treated as tangible and intangible cultural heritage subject to safeguarding, protection, and promotion, as well as to administrative procedures for their identification, delimitation, demarcation, and titling. Both proposals advocate, therefore, that the use of the coastal-marine territory, including its various ecosystems such as rocky shorelines, should be guaranteed, within environmentally responsible standards, for the exercise of socioeconomic activities by vulnerable social groups, such as artisanal fishermen and gatherers.

Similar legislation exists or is under consideration in other countries, such as in Australia, where a fisheries co-management policy consistent with the Fisheries Management Act of 1991 has been adopted as a means to encourage adaptation.

[1] Other problems emerging with global sea warming are associated with health, such as the proliferation of Harmful Algal Blooms (HABs) and the spread of *Vibrio cholera* virus in vectors stimulated by temperature increase. The increasing likelihood of occurrences of HABs under climate change also elevates their risks on ecosystem services such as fisheries, aquaculture and tourism as well as public health (Bindoff et al. 2019).

[2] https://www.camara.leg.br/proposicoesWeb/fichadetramitacao?idProposicao=604557

[3] https://www.camara.leg.br/proposicoesWeb/fichadetramitacao?idProposicao=2236610

Existing legislation as part of Brazil's national system of protected areas (SNUC Law 9.985/2000[4]) provides for demarcation of Marine Extractive Reserves (RESEX--Mar), currently 16 in number, which have been created throughout Brazil's coastal zone including areas such as Arraial do Cabo in northern Rio de Janeiro, where fishers depend primarily on rocky shorelines. Such reserves involve participatory co-management of marine artisanal fisheries between fisher communities and state or federal fishing and conservation agencies to protect marine and coastal resources and to provide legitimacy to the occupation and use of the coastal territory for productive purposes in accordance with approved reserve management plans.

As Seixas and Kalikoski (2009), Oliveira et al. (2018), Lopes et al. (2015), and Mattos and Wojciechowski (2019), among others, conclude, the experiences of agreements on the rights of use and access to natural resources relying on participatory processes of co-management of coastal-marine territory must be continually strengthened, to perpetuate protection of rocky shore and intertidal marine ecosystems and the human populations that depend upon them.

The Cultural Importance of Rocky Shores: More Than Just Coastlines

Recreational and Touristic Opportunities

Rocky shores offer an extraordinary array of recreational opportunities that extend beyond their aesthetic appeal. The combination of unique geological formations (Bacci et al. 2019), diverse marine life, and striking seascapes creates a compelling destination for various leisure activities, while also contributing to physical and mental health.

Firstly, rocky shores provide opportunities for low-impact recreational activities such as tide pooling, subaquatic interpretive trails, hiking, and birdwatching. According to Bertocci et al. (2012), rocky intertidal zones serve as a natural classroom for curious visitors to observe tide pools filled with starfish, anemones, and various crustaceans. This hands-on experience is both educational (Pedrini et al. 2019; Pimentel et al. 2019) and therapeutic, promoting a deeper connection with nature and reinforcing the importance of marine biodiversity conservation (Wyles 2014).

Additionally, rocky shores offer the opportunity for adventurous activities like snorkeling, rock climbing, and scuba diving, attracting outdoor enthusiasts seeking challenges in stunning landscapes. The diversity of underwater habitats and species seen in these areas often provides an authentic and immersive experience (Thompson et al. 2002; Berchez et al. 2007) that can significantly contribute to local economies through ecotourism, which has grown in popularity among divers and nature

[4] https://www.planalto.gov.br/ccivil_03/leis/l9985.htm

enthusiasts. Several studies have measured the impacts of educational activities on rocky shores (Berchez et al. 2005; Pedrini et al. 2008), which is a crucial step towards sustainable tourism in these areas. The effective management of recreational activities can benefit both biodiversity and community livelihoods (Silva and Ghilardi-Lopes 2012; Pedrini et al. 2016).

Educational Value

Rocky shores provide invaluable opportunities for education in various disciplines, from marine biology to environmental conservation. Their unique ecological characteristics offer students, educators, and researchers an opportunity for experiential learning about all seven principles of Ocean Literacy, inside and outside school spaces (Ghilardi-Lopes et al. 2012, 2019, 2023).

Firstly, rocky shores are ideal for understanding species diversity and adaptations. Intertidal zones are marked by stark environmental gradients that expose organisms to varying temperatures, salinity, and wave action. According to Raffaelli and Hawkins (2012), the distribution of species in intertidal zones can illustrate ecological concepts such as competition, predation, and niche differentiation, making rocky shores an outdoor classroom.

Secondly, they provide a natural laboratory for understanding the impacts of environmental change (Berchez et al. 2016). Due to their proximity to human populations and sensitivity to global climate shifts, rocky shores are particularly vulnerable to pollution, ocean acidification, and rising sea levels (Coutinho et al. 2016). The impacts of these changes are readily observable, allowing students to directly connect theory with practice. This firsthand observation fosters environmental stewardship and emphasizes the urgency of conservation efforts. Furthermore, rocky shores serve as platforms for community-based education, often involving citizen science initiatives that engage the public in data collection and monitoring. For example, projects like the Shorewatch program (UK), "De Olho na Costa" (Brazil) and "Active Citizens" (Brazil) have empowered local communities to monitor marine biodiversity (Kasten et al. 2021; Lacerda et al. 2023) and several initiatives around the world empowered people to monitor marine litter (Kawabe et al. 2022). This inclusion creates a broader understanding of the ecological value of rocky shores, helping translate scientific research into actionable conservation strategies.

Cultural Heritage and Artistic Inspiration

Coastal communities have long been intertwined with rocky shores. Many coastal traditions, folklore, and artistic expressions find their roots in these landscapes. Artists, writers, and photographers draw inspiration from the shore's vivid contrast between rock formations, marine life, and ocean waves, while local traditions, such as fishing or gathering shellfish, continue to be important cultural practices.

Artists and photographers, such as Roberto Burle Marx and Sebastião Salgado, have been inspired by the dramatic interplay of rock formations and marine life, capturing the rugged beauty of the shoreline.

Pedra do Sal, for example, located in the Saúde neighborhood of Rio de Janeiro, Brazil, is historically significant as a cultural hub for Afro-Brazilian communities and the birthplace of Samba music. This site was originally a gathering place for African slaves, particularly those from Bahia, where they engaged in cultural activities and laid the groundwork for what would later become Samba. This musical genre, deeply rooted in African rhythms and traditions, emerged and evolved in the informal gatherings held at Pedra do Sal. The site is now recognized for its vibrant street music scene, where locals and tourists alike come to experience live performances, often featuring samba and other Afro-Brazilian genres (Guimarães 2013).

Also, coastal landscapes have influenced regional literature, with the incorporation of rocky shore imagery to symbolize resilience and the interconnectedness of nature and people. In 2021, the Brazilian Institute of Geography and Statistics (IBGE) published an Atlas that portrays literary representations of Brazilian coastal regions that were shaped by the dynamics of occupation, confrontations, and alliances between people, the slave trade and slavery. The regions cited in Brazilian literature illustrate the great diversity of one of the longest—more than 8000 km—and most inhabited coastal strips in the world, characterized not only by economic cycles or extraction of natural resources, but by their role as an interface between the ocean and the continent, between the Old and New Worlds. Several regions of historical importance are represented in this literature: the cocoa region of Southern Bahia, the Recôncavo areas of Salvador and Recôncavo Baiano (state of Bahia); Santos city and the state of São Paulo Coast; the sugar hinterland in Olinda and Recife (state of Pernambuco) and Costa dos Engenhos (several Northeastern States); Rio de Janeiro city, Norte Fluminense and Guanabara Bay (state of Rio de Janeiro); the recesses of São Luís and Baía de São Marcos (state of Maranhão); Belém and Foz do Rio Amazonas (state of Pará); Rio Grande de São Pedro and Costa Sul (state of Rio Grande do Sul); and Santa Catarina Island (state of Santa Catarina).

Historical Traditions and Indigenous Cultures

Rocky shores have played a significant role in the historical and cultural development of coastal communities. Many indigenous cultures, such as the Tupinambá and Caeté, utilized rocky intertidal zones for gathering shellfish and small marine organisms. These practices, which shaped the traditional diets of indigenous groups, continue to be practiced in modified forms by modern communities (Mourão et al. 2020), emphasizing the intergenerational knowledge passed down through oral traditions (Silva et al. 2022).

Traditional rafts, or "jangadas," hold a significant place in Brazil's coastal history. These watercrafts, constructed from various local wood species and bound logs, have been central to artisanal fishing traditions for generations. In Bahia, fishermen utilize their ecological knowledge to craft rafts using specific wood types like *Apeiba tibourbou* and *Eschweilera ovata*, chosen for their buoyancy and flexibility (Andrade et al. 2018). In Ilha da Maré, the artisanal fishing traditions of a traditional quilombola community remain a central part of their identity (Sales and Vidal 2020).

Spiritual and Mental Well-Being

Beyond economic contributions, rocky shores also positively impact mental and physical health, which can result in positive change in attitudes and behaviors towards the environment (Wyles and Ghilardi-Lopes 2023). The restorative power of natural environments has been well-documented. White et al. (2010) and Severin et al. (2023) found that exposure to coastal landscapes, particularly rocky shores, helps reduce stress and promotes overall well-being. Such benefits make these environments indispensable for urban populations seeking natural respite.

References

Acheson JM (1981) Anthropology of fishing. Annu Rev Anthropol 10:275–316

Andrade ILMM, Mielke MS, Peroni N, Schiavetti A (2018) Fishermen do more than fish: local ecological knowledge of raftsmen about the arboreal species used to construct rafts (Bahia, Brazil). J Ethnobiol Ethnomed 14(80). https://doi.org/10.1186/S13002-018-0279-7

Bacci DC, Garcia MGM, Bourotte CLM, Santos VMN (2019) Integration of marine and coastal geo-biodiversity in environmental education through earth system science and social learning approaches. In: Ghilardi-Lopes N, Berchez F (eds) Coastal and marine environmental education. Brazilian marine biodiversity, chapter 3. Springer, Cham, pp 39–52. https://doi.org/10.1007/978-3-030-05138-9_3

Bayley PB, Petrere M Jr (1989) Amazon fisheries: assessment methods, current status and management options. Can Spec Publ Fish Aquat Sci 106:385–398

Begossi A (1995) Fishing spots and sea tenure: incipient forms of local management in Atlantic Forest coastal communities. Hum Ecol 23(3):387–406. https://doi.org/10.1007/BF01190138

Begossi A (2006) Temporal stability in fishing spots: conservation and co-management in Brazilian artisanal coastal fisheries. Ecol Soc 11(1):5–25. https://doi.org/10.5751/ES-01380-110105

Begossi A, Camargo E, Capri S (2013) Os mapas da pesca artesanal – pesqueiros e pescadores na costa do Brasil. Rima Editora, São Carlos, 166 p

Berchez FAS, Carvalhal F, Robim MJ (2005) Underwater interpretative trail: guidance to improve education and decrease ecological damage. Int J Environ Sustain Dev 4(2):128–139. https://doi.org/10.1504/IJESD.2005.007235

Berchez F, Ghilardi N, Robim MJ, Pedrini AG, Hadel VF, Fluckiger G, Simões M, Mazzaro R, Klausener C, Sanches C, Bespalec P (2007) Projeto Trilha Subaquática: sugestão de diretrizes para a criação de modelos de educação ambiental em unidades de conservação ligadas a ecossistemas marinhos. OLAM-Ciênc Tecnol 7(3):181–209

Berchez FAS, Ghilardi-Lopes NP, Correia MD, Sovierzoski HH, Pedrini AG, Ursi S, Kremer LP, Almeida R, Schaeffer-Novelli Y, Marques V, Brotto DS (2016) Marine and coastal environmental education in the context of global climate changes – synthesis and subsidies for ReBentos (Coastal Benthic Habitats Monitoring Network). Braz J Oceanogr 64(sp. 2):137–156. https://doi.org/10.1590/S1679-875920160932064sp2

Berkes F, Colding J, Folke C (2000) Rediscovery of traditional ecological knowledge as adaptive management. Ecol Appl 10:1251–1262

Bertocci I, Dominguez JH, Freitas C, Sousa-Pinto I (2012) Patterns of variation of intertidal species of commercial interest in the Parque Litoral Norte (North Portugal) MPA: comparison with three reference shores. Mar Environ Res 77:60–70. https://doi.org/10.1016/j.marenvres.2012.02.003

Bindoff NL, Cheung WW, Kairo JG, Aristegui J, Guinder VA, Hallberg R, ... & Williamson P (2019) Changing ocean, marine ecosystems, and dependent communities. Cambridge University Press, pp 447–587. https://ri.conicet.gov.ar/handle/11336/233328

Clark G, Fischer M, Hunter C (2021) Coasts. Beaches and shorelines. In: Cresswell I, Janke T, Johnston EL (eds) Australia. State of the Environment (2021). https://soe.dcceew.gov.au/coasts/environment/beaches-and-shorelines

Clauzet M (2009) Etnoicitiologia de pescadores artesanais do nordeste e sudeste do Brasil. Tese de Doutorado. Núcleo de Estudos e Pesquisas Ambientais, Universidade Estadual de Campinas, Campinas, São Paulo, 175 p. https://doi.org/10.47749/T/UNICAMP.2009.471669

Clauzet M, Ramires M, Begossi A (2007) Ethnoichthyology of artisanal fishing community from Guaibim Beach, Valença (BA), Brazil. Neotrop Biol Conserv 2(3):136–154

Coutinho R, Yaginuma LE, Siviero F, Santos JCQ, López MS, Christofoletti RA, Berchez F, Ghilardi-Lopes NP, Ferreira CEL, Gonçalves JEA, Masi BP, Correia MD, Sovierzoski HH, Skinner LF, Zalmon IR (2016) Studies on benthic communities of rocky shores on the Brazilian coast and climate change monitoring: status of knowledge and challenges. Braz J Oceanogr 64(sp2):27–36. https://doi.org/10.1590/S1679-875920161015064sp2

Crowe TP, Thompson RC, Bray S, Hawkins SJ (2000) Impacts of anthropogenic stress on rocky intertidal communities. J Aquat Ecosyst Stress Recover 7:273–297. https://doi.org/10.1023/A:1009911928100

Diegues ACS (1983) Pescadores, camponeses e trabalhadores do mar. Ensaios 94, Ática, São Paulo, 287 p

Diegues ACS (1995) Povos e mares: uma retrospectiva de sócio-antropologia marítima. NUPAUB--USP, São Paulo, 269 p

Elliff CI, Kikuchi RKP (2015) The ecosystem service approach and its application as a tool for integrated coastal management. Perspect Ecol Conserv 13(2):105–111. https://doi.org/10.1016/j.ncon.2015.10.001

FAO (2015) Voluntary guidelines for securing sustainable small-scale fisheries in the context of food security & poverty eradication. FAO, Rome. https://www.fao.org/voluntary-guidelines-small-scale-fisheries/en

Garcia MGM (2019) Ecosystem services provided by geodiversity: preliminary assessment and perspectives for the sustainable use of natural resources in the coastal region of the State of São Paulo, Southeastern Brazil. Geoheritage 11:1257–1266. https://doi.org/10.1007/s12371-019-00383-0

Garcia MGM, Brilha J, Souza CRG, Del Lama EA (2018) Preliminary assessment of ecosystem services provided by geodiversity in the coastal region of the state of São Paulo, Southeastern Brazil. In: Głowniak E, Wasiłowska A, Leonowicz P (eds) Geoheritage and conservation: modern approaches and applications towards the 2030 agenda. 9th ProGEO symposium, Chęciny, Poland, 25–28th June 2018 Programme and abstract book, pp 101–102

Ghilardi-Lopes NP, Hadel VF, Berchez FAS (2012) Guia para educação ambiental em costões rochosos. Artmed, São Paulo, 200 p

Ghilardi-Lopes NP, Kremer LP, Barradas JI (2019) The importance of "ocean literacy" in the Anthropocene and how environmental education can help in its promotion. In: Ghilardi-Lopes N, Berchez F (eds) Coastal and marine environmental education. Brazilian marine biodiversity, chapter 1. Springer, Cham, pp 3–17. https://doi.org/10.1007/978-3-030-05138-9_1

Ghilardi-Lopes NP, Motokane M, Barradas JI, Xavier L, Menck EVS, Franco ACG, Turra A (2023) Ocean as an interdisciplinary theme in Brazilian basic education. Ambiente Soc 26:1–22. https://doi.org/10.1590/1809-4422asoc20210134vu2023L2OA

Guimarães RS (2013) Urban interventions, memories and conflicts: black heritage and the revitalization of Rio de Janeiro's Port Zone. Vibrant: Virtual Braz Anthr 10:208–227. https://doi.org/10.1590/S1809-43412013000100011

Hadley D (2009) Land use and the coastal zone. Land Use Policy 26(S1):S198–S203. https://doi.org/10.1016/j.landusepol.2009.09.014

IBGE (2021) Atlas das representações literárias de regiões brasileiras. Volume 4: Costa Brasileira. IBGE, Rio de Janeiro, 2006–2021, 298 p

Intergovernmental Panel on Climate Change – IPCC (2022) Changing ocean, marine ecosystems, and dependent communities. In: The ocean and cryosphere in a changing climate: special report of the intergovernmental panel on climate change, chapter 5. Cambridge University Press, Cambridge, pp 447–588. https://doi.org/10.1017/9781009157964.007

Johannes RE, Freemen MMR, Hamilton JR (2000) Ignore fishers' knowledge and miss the boat. Fish Fish 1:257–271. https://doi.org/10.1111/j.1467-2979.2000.00019.x

Kasten P, Jenkins SR, Christofoletti RA (2021) Participatory monitoring—a citizen science approach for coastal environments. Front Mar Sci 8:681969. https://doi.org/10.3389/fmars.2021.681969

Kawabe LA, Ghilardi-Lopes NP, Turra A, Wyles KJ (2022) Citizen science in marine litter research: a review. Mar Pollut Bull 182:114011. https://doi.org/10.1016/j.marpolbul.2022.114011

Lacerda JVA, Koffler S, Gonzalez JD, Monteiro GB, Kawabe LA, Mendes M, Bravo P, Ghilardi-Lopes NP (2023) Many possibilities of students as citizen scientists: a supplement to Forti (2023). Biodiversity 24:1–5. https://doi.org/10.1080/14888386.2023.2256712

Lopes PFM, Pacheco S, Clauzet M, Silvano RAM, Begossi A (2015) Fisheries, tourism, and marine protected areas: conflicting or synergistic interactions? Ecosyst Serv 16:333–340. https://doi.org/10.1016/j.ecoser.2014.12.003

Mattos SMG, Wojciechowski MJ (2019) Existing institutional and legal framework and its implications for small-scale fisheries development in Brazil. In: Salas S, Barragán-Paladines MJ, Chuenpagdee R (eds) Viability and sustainability of small-scale fisheries in Latin America and the Caribbean. MARE Publication Series 19. Springer, Heidelberg, pp 495–511. https://doi.org/10.1007/978-3-319-76078-0_21

Mattos SMG, Woiciechowski MJ, Gandini FC (2020) Iluminando as capturas ocultas da pesca artesanal costeira no Brasil: Um estudo de caso. https://spark.adobe.com/page/tONM9fbNtuvuj/

Mendonça JT, Castro PMGC, Cabral IM, Carvalho MES (2018) Emprego de métodos participativos, qualitativos e mistos na pesquisa voltada para a gestão pesqueira no Brasil. métodos participativos, qualitativos e mistos na pesquisa voltada para a gestão pesqueira no Brasil. In: Brandão C, Carvalho JL, Ribeiro J, Costa AP (eds) A prática na investigação qualitativa: exemplos de estudos, vol 2, Ludomedia, Aveiro, pp 55–90

Mearns AJ, Morrison AM, Arthur C, Rutherford N, Bissell M, & Rempel-Hester MA (2020) Effects of pollution on marine organisms. Water Environ Res 92(10):1510–1532. https://doi.org/10.1002/wer.1400

Mieszkowska N, Burrows MT, Hawkins SJ, Sugden H (2021) Impacts of pervasive climate change and extreme events on rocky intertidal communities: evidence from long-term data. Front Mar Sci 8:642764. https://doi.org/10.3389/fmars.2021.642764

Mourão JS, Baracho RL, Martel G, Barboza RRD, Lopes SF (2020) Local ecological knowledge of shellfish collectors in an extractivist reserve, Northeast Brazil: implications for co-management. Hydrobiologia 847:1977–1997. https://doi.org/10.1007/s10750-020-04226-w

Oliveira C, Marenzi CR, Longarete C (2018) Serviços ecossistêmicos prestados pelos promontórios costeiros no litoral centro-norte de Santa Catarina. Braz J Aquat Sci Technol 21(2):10–18. https://doi.org/10.14210/bjast.v21n2.11172

Pedrini AG, Costa C, Silva VG, Maneschy FS, Newton T, Berchez F, Ghilardi NP, Spelta L (2008) Gestão de áreas protegidas e efeitos da visitação ecoturística pelo mergulho com snorkel: O caso do parque estadual da Ilha Anchieta (Peia), Estado de São Paulo, Brasil. REMEA-Rev Eletrônica Mestr Educ Ambient 20:1–20

Pedrini AG, Brotto DS, Pimentel DS, Behrends E, de Moraes AAJ (2016) Transformative and Emancipatory Environmental Education by Marine Ecotourism in the marine environmental protection area of Armação dos Búzios (RJ, Brazil). Rev Bras Ecoturismo 9(3):445–470. https://doi.org/10.34024/rbecotur.2016.v9.6531

Pedrini AG, Kremer LP, Brotto DS, Silva VRF (2019) Emblematic coastal and marine environmental education projects in Brazil. In: Ghilardi-Lopes N, Berchez F (eds) Coastal and marine environmental education. Brazilian marine biodiversity, chapter 6. Springer, Cham, pp 87–101. https://doi.org/10.1007/978-3-030-05138-9_6

Pimentel DS, Pedrini AG, Kawabe LA, Robim MJ, Berchez FAS, Meireles CP (2019) Opportunities and challenges to implement environmental education in Brazilian coastal and marine protected areas. In: Ghilardi-Lopes N, Berchez F (eds) Coastal and marine environmental education. Brazilian marine biodiversity, chapter 7. Springer, Cham, pp 103–117. https://doi.org/10.1007/978-3-030-05138-9_7

Raffaelli D, Hawkins S (2012) Intertidal ecology. Springer Science & Business Media, 356 p. https://doi.org/10.1007/978-94-009-1489-6

Sales IG, Vidal JP (2020) A comunidade tradicional quilombola de pescadores artesanais e marisqueiras da Ilha da Maré, no Brasil, e sua exclusão social, histórica e cultural: um processo de invisibilidade. Rev Int de Educ Tecnol 1(1). https://doi.org/10.30612/riet.v1i1.12955

Seixas CS, Kalikoski D (2009) Gestão participativa da pesca no Brasil: Levantamento das iniciativas e documentação dos processos. Desenvolv Meio Ambiente 20:119–139. https://doi.org/10.5380/dma.v20i0.12729

Severin MI, Akpetou LK, Annasawmy P, Asuquo FE, Beckman F, Benomar M, Jaya-Ram A, Malouli M, Mees J, Monteiro I, Ndwiga J, Neves Silva P, Nubi OA, Sim YK, Sohou Z, Shau-Hwai AT, Woo SP, Zizah S, Buysse A, Raes F, Krug LA, Seeyave S, Everaert G, Mahu E, Catarino AI (2023) Impact of the citizen science project COLLECT on ocean literacy and well-being within a north/west African and south-east Asian context. Front Psychol 14:1130596. https://doi.org/10.3389/fpsyg.2023.1130596

Silva JN, Ghilardi-Lopes NP (2012) Indicators of the impacts of tourism on hard-bottom benthic communities of Ilha do Cardoso State Park (Cananéia) and Sonho Beach (Itanhaém), two southern coastal areas of São Paulo State (Brazil). Ocean Coast Manag 58:1–8. https://doi.org/10.1016/j.ocecoaman.2011.12.009

Silva GS, Oliveira FP, Silva JB, Barros SHB (2022) Generational knowledge and learning in the productive practices of traditional coastal communities, Brazilian Amazon. Int J Innov Educ Res 10(3):189–207. https://doi.org/10.31686/ijier.vol10.iss3.3679

Starko S, Bailey LA, Creviston E, James KA, Warren A, Brophy MK, Danasel A, Fass MP, Townsend JA, Neufeld CJ (2019) Environmental heterogeneity mediates scale-dependent declines in kelp diversity on intertidal rocky shores. PLoS One 14(3):e0213191. https://doi.org/10.1371/journal.pone.0213191

Thompson RC, Crowe TP, Hawkins SJ (2002) Rocky intertidal communities: past environmental changes, present status and predictions for the next 25 years. Environ Conserv 29(2):168–191

UNEP (2006) Marine and coastal ecosystems and human wellbeing: a synthesis report based on the findings of the Millennium Ecosystem Assessment. UNEP, 76 p. https://www.millenniumassessment.org/documents/Document.799.aspx.pdf

White MP, Smith A, Humphryes K, Pahl S, Snelling D, Depledge MH (2010) Blue space: the importance of water for preference, affect, and restorativeness ratings of natural and built scenes. J Environ Psychol 30(4):482–493. https://doi.org/10.1016/j.jenvp.2010.04.004

Wyles KJ (2014) Rocky shores: from habitat threat to marine awareness & well-being benefits. Doctoral dissertation, Plymouth University

Wyles K, Ghilardi-Lopes NP (2023) Citizen science as a pro-environmental behaviour and a catalyst for further behaviour change. In: Gatersleben B, Murtagh N (eds) Handbook on pro-environmental behaviour change, chapter 21. Edward Elgar Publishing, UK, pp 321–334

Chapter 6
Anthropogenic Stressors on Shallow Coastal Rocky Reefs

Aline Sbizera Martinez and Luciana Altvater

Abstract In this chapter, we provide an overview of the impacts of human activities on Brazilian rocky reefs. We highlight the increase in population and urbanization, which have amplified pressures such as overfishing, pollution, habitat destruction, and climate change on these reefs. It is important to note that while excessive artisanal fishing has led to reduced fish stocks, the broader impacts on invertebrate species remain largely unknown. Moreover, land and sea activities, together with the widespread issue of untreated sewage along the Brazilian Coast, pose significant threats to marine life, affecting various species and benthic assemblages. Additionally, coastal urban development and recreational activities contribute to habitat loss, shifts in population densities and assemblage's structure. Marine bioinvasion, driven by increasing global trade and maritime shipping, the offshore industry and coastal urban sprawl, has led to reductions in local native biodiversity and alterations in community structure and composition. These changes can be exacerbated by the effects of ocean warming, acidification, and rising sea levels, as shown for a few invertebrate species and algal assemblages. Efforts by Brazilian researchers to better guide sustainable coastal management should address local scientific gaps, given the spatial fragmentation of current research. Such comprehensive knowledge is crucial for mitigating anthropogenic stressors and ensuring the sustainability of Brazil's coastal rocky reefs.

Supplementary Information The online version contains supplementary material available at https://doi.org/10.1007/978-3-031-67206-4_6.

A. S. Martinez (✉)
Instituto do Mar, Universidade Federal de São Paulo, Santos, São Paulo, Brazil

L. Altvater
Departamento de Biotecnologia Marinha, Instituto de Estudos do Mar Almirante Paulo Moreira, Arraial do Cabo, Rio de Janeiro, Brazil

R. Coutinho, R. A. Christofoletti (eds.), *Brazilian Rocky Shores*, Brazilian Marine Biodiversity, https://doi.org/10.1007/978-3-031-67206-4_6

Introduction

Humans have a long-time relationship with the sea, using its resources for food, living and seafaring. Early records of marine resources exploitation with impacts on nearshore ecosystems are dated back in the late Pleistocene era (Erlandson and Rick 2009). As the human population expanded, so did the magnitude of our footprint on natural systems. We now live in a world where virtually all ecosystems are impacted by mankind to a greater or lesser extent (Vitousek et al. 1997; Sanderson et al. 2002). Fishing has historically preceded all human pressures on marine ecosystems, which was followed by increased pollution, habitat modification, bioinvasion and climate change, as population has grown (Jackson et al. 2001).

Local disturbances on the Brazilian Coast have initially concentrated around estuarine ecosystems, where ports, industries, urban areas, and fishery processing and trade first developed (Borelli 2007; Moraes 2007; Fistarol et al. 2015; Mattos et al. 2017). Port activities started during the colonization period for transportation of raw material to Europe, a time when fishing was primarily for sustenance of indigenous people (Moraes 2007; Instituto da Pesca 2017). Commercial fishing of prawns and sardine for exportation initiated early, in the twentieth century (Instituto da Pesca 2017). From the 1950s, Brazil went through a heavy industrialization process, which also advanced fishing technology and boosted fishery production (Moraes 2007, Instituto da Pesca 2017). Since the 1980s, urban areas have expanded into marine coastal areas, a growth accelerated by population migration to coastal cities. More recently, this expansion has been further fuelled by tourism development (Moraes 2007).

Understanding the stressors associated with the history of land occupation and resource use on the Brazilian coastline is crucial for drawing research directions and prioritizing management actions that will maintain the provision of ecosystem services. In this chapter, we reviewed the current state of knowledge concerning the existing stressors (fishing, coastal contamination, habitat modification, bioinvasion and climate change) and known impacts on shallow rocky reefs along the Brazilian Coast that are linked to human activities.

Extractive Fishery and Aquaculture

Brazilian fisheries currently rank the 18th position in worldwide fishery production (Mattos et al. 2020). Seventy percent of marine extractive fisheries contribute to the national exploitative production (MPA/Brasil 2010). Nearly 87% of this production probably comes from artisanal fisheries, i.e. small-scale fisheries (Mattos et al. 2020). This information is important since much of the living resource exploitation on rocky reefs is related to small-scale fisheries, either by hand picking, spearfishing, or the use of small fishing traps, nets and boats (e.g. Lage and Jablonski 2008; Nunes et al. 2012; Marinho-Soriano 2017; Silvano et al. 2017). A recent review indicates that marine fisheries in Brazil have led to population declines in numerous fish species, with artisanal fisheries contributing to the majority of these catches (Eggertsen et al. 2024). Nonetheless, accurate data on the artisanal fishery production of rocky reef species, particularly invertebrates, remains limited and fragmented due to the absence of monitoring programmes that systematically gather information on these fisheries (Neto et al. 2021; Eggertsen et al. 2024).

Harvesting of Rocky Reef Fauna and Flora

In this section, we explore the diverse array of species that inhabit or are associated with Brazil's shallow rocky shores, focusing on their fisheries use, conservation status, and legal frameworks governing their harvest. This analysis does not cover catch numbers, which were comprehensively reviewed by Eggertsen et al. (2024), but it provides an understanding of the broader socio-ecological and regulatory contexts.

There are about 425 species that are harvested from shallow reefs in Brazil among fishes (244 spp.), invertebrates (154 spp.), and macroalgae (27 spp.) (Fig. 6.1; Supplementary Material 6.1). Numerous species have been harvested for various purposes, with the majority (60%–328 species: 191 fish, 117 invertebrates, and 20 algae) targeted primarily for the aquarium trade (Fig. 6.1). These include many ornamental fishes, molluscs (e.g. snails and clams), crustaceans (e.g. shrimps and crabs), echinoderms (e.g. sea stars and sea urchins), cnidarians (e.g. corals and sea anemones), sponges, polychaetes, ascidians and macroalgae (Supplementary Material 6.1).

Brazil is one of the main suppliers of marine organisms for the worldwide aquarium market, whose harvesting activities on natural reefs have boosted in the 1990s (Nottingham et al. 2000; Monteiro-Neto et al. 2003; Gasparini et al. 2005; Wood 2001). However, information on the amounts harvested from the reefs and traded in both the international and national markets is still inaccurate (Gasparini et al. 2005; Gurjão et al. 2017; Gurjão and Lotufo 2018). This is a major threat to many rocky reef organisms since basic knowledge on natural abundances, population dynamics of many species and diversity patterns are scarce (Monteiro-Neto et al. 2003; Araújo et al. 2020). Moreover, numerous species are frequently harvested for a variety of uses, exacerbating the fishing pressure on them (Supplementary Material 6.1).

About 27% of the harvested fish (93 spp.) are for consumption (Fig. 6.1), including targeted and bycatch species (Supplementary Material 6.1). This fishery is characterized by artisanal activities using gill nets, traps, seine, trolling and angling either for subsistence or commercial purposes and by recreational fishing, often angling and spearfishing (Frédou et al. 2006; Pinheiro et al. 2010). Approximately 9% (30 spp.) of the species that are caught for consumption are discarded (Fig. 6.1). The intensive fishing activity has severely depleted fish populations, placing some species at risk of extinction (Eggertsen et al. 2024). Despite their conservation status, these species are still subject to illegal fishing for consumption (Giglio et al.

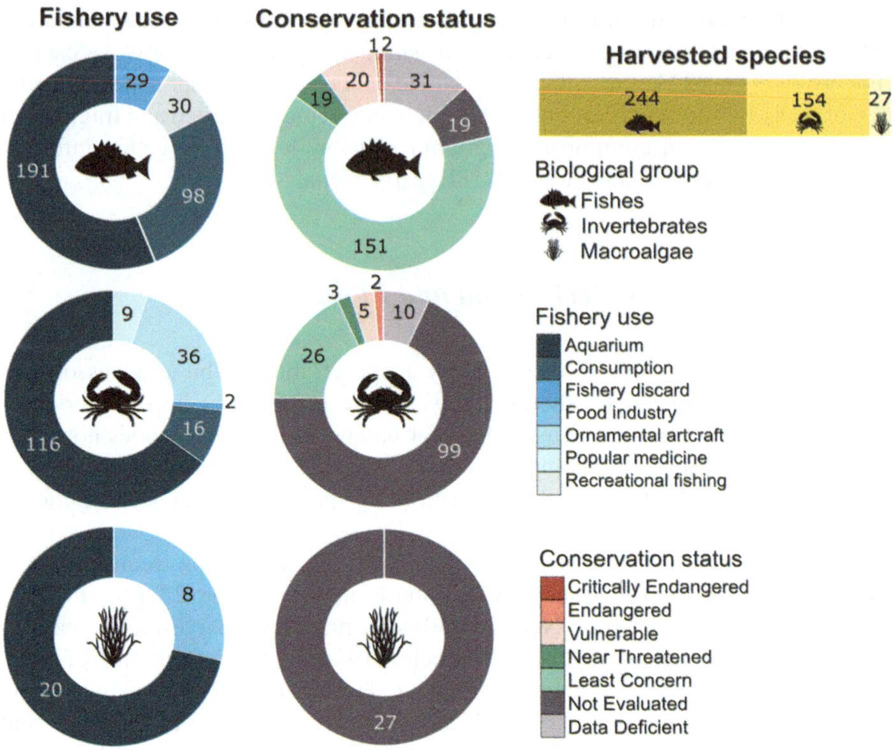

Fig. 6.1 Overview of species harvested from shallow reefs of Brazil with detailed information on the number of taxonomic groups of fish, invertebrates and macroalgae harvested for different uses and their conservation statuses. Conservation statuses were quantified according to the 2018 Brazilian list of threatened fauna (for fishes and invertebrates) and the decree MMA 443/2014 (for macroalgae), which included the categories: Not Evaluated (NE), Data Deficient (DD), Least Concern (LC), Near Threatened (NT), Vulnerable (VU), Endangered (EN), Critically Endangered (CR). The graphs used in this figure were generated with R version 4.3.2 and subsequently edited and assembled in Inkscape version 1.3.2. (Icons were sourced from SVG Repo and the IAN Symbol Libraries)

2017, 2018; Pereira et al. 2021). Nearly 10% (23 spp.) of the fishes are under threat of extinction according to the 2018 Brazilian list of threatened fauna (ICMBio/ MMA 2018), and fishing is thereby forbidden by law. Of these, 20 species are vulnerable, 2 species are endangered (the grouper *Epinephelus marginatus* (Lowe, 1834) and the parrotfish *Scarus trispinosus* Valenciennes, 1840), and 1 species (the grouper *Epinephelus itajara* (Lichtenstein, 1822) is critically endangered (Fig. 6.1, Supplementary Material 6.1)). Another 20% of the fishes are also illegally fished for consumption (Fig. 6.2), of which 7% are specifically prohibited for the aquarium trade under the regulating norms IN Ibama 202/2008 and decree MMA 445/2014.

Although fishing is permitted for 70% of fish species in Brazil, and the conservation status of 43% is classified as "least concern", populations of some species have experienced declines at local scales (Gasparini et al. 2005; Pinheiro et al. 2010;

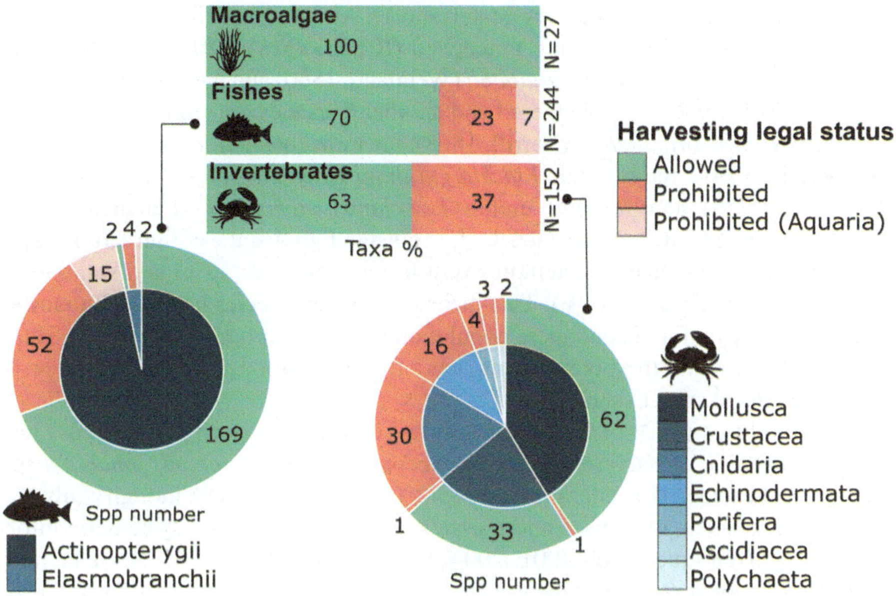

Fig. 6.2 Harvesting legal status (%) of species harvested from shallow reefs of Brazil with the detailed information on the taxonomic groups (number of spp.) of fish and invertebrates. Harvesting legal status was quantified according to the regulating norms IN Ibama 202/2008 (for fishes), federal law 9605/2008 (for invertebrates and macroalgae) and the decrees MMA 445/2014 (for fishes and invertebrates) and MMA 443/2014 (for macroalgae). The graphs used in this figure were generated with R version 4.3.2 and subsequently edited and assembled in Inkscape version 1.3.2. (Icons were sourced from SVG Repo and the IAN Symbol Libraries)

Silvano et al. 2017; Guabiroba et al. 2020; Eggertsen et al. 2024). Furthermore, the intense fishing pressure has changed the size of target fishes and the structure of fish assemblages (Ferreira and Gonçalves 1999; Gasparini et al. 2005; Floeter et al. 2006; Frédou et al. 2006; Eggertsen et al. 2024). Target large predators and herbivores are often bigger and more abundant within non-take zones, whereas non-target species are more abundant outside of protected areas (Ferreira and Gonçalves 1999; Floeter et al. 2006; Anderson et al. 2014; Rolim et al. 2019).

Fish harvesting has also caused indirect effects on sessile benthic organisms. For instance, the methods used to collect fish for the aquarium trade impact invertebrates, algae and associated fauna through the destruction of benthic habitats (Nottingham et al. 2000, 2005; Wood 2001; Gasparini et al. 2005). Another indirect impact is ghost fishing, which results from the loss or illegal discard of fishing waste that ends up on the rocky surface killing many reef associated organisms, such as invertebrates, fish and sea turtles (Adelir-Alves et al. 2016; Link et al. 2019).

Invertebrates face threats not only from the indirect impacts of fish harvesting but also from direct exploitation. Of the 117 invertebrate species harvested from Brazilian reefs, 37% are for the aquarium trade, 12% to the commerce of ornamental art craft and souvenir, 5% are for consumption, and 3% are used for medicinal purposes (Fig. 6.1, Supplementary Material 6.1). About 5% of the harvested invertebrates (i.e. 7 spp., Fig. 6.1) are threatened (ICMBio/MMA 2018). Of these, the giant sea anemone *Condylactis gigantea* (Weinland, 1860) and the coral *Mussismilia harttii* (Verrill, 1868) are endangered and another 5 species are vulnerable, viz.: the coral *Mussismilia braziliensis* (Verrill, 1868), the milk conch *Macrostrombus costatus* (Gmelin, 1791), the sea stars *Linckia guildingi* Gray, 1840, *Oreaster reticulatus* (Linnaeus, 1758) and the sea urchin *Lytechinus variegatus* (Lamarck, 1816) (Fig. 6.1, Supplementary Material 6.1). It is outstanding that the conservation status of most of the harvested invertebrates is still not evaluated (70%) and some were classified as data deficient (7%). This means that many species lack baseline information on distribution, abundance, population traits or reproductive biology. Thus, it is hard to evaluate the potential impacts of harvesting on the populations or to properly manage such fisheries.

Only molluscs (43% of harvested species, except the conch *M. costatus*) and crustaceans (23% of harvested species, except the shrimp *Thor* aff. *amboinensis*) are permitted under the federal law 9605/2008 (Fig. 6.2). Even though, only the fisheries of lobsters, octopuses and mussels have relatively more systematic information (e.g. Henriques et al. 2001, 2004; Lage and Jablonski 2008; Santana et al. 2010; Haimovici et al. 2014b; Duarte et al. 2016; Silva and Martins 2017), and fishing activities are managed by detailed regulating norms, viz. IN SEAP/PR 15/2007, IN Ibama 138/2006, IN Ibama 105/2006, IN Ibama 54/2019. In many cases, however, fishermen do not adhere to regulations by fishing specimens during the protection period and catching amounts above the established limit or individuals below the size permitted by law (Aragão 2013). Beyond legal considerations, declines in octopus (since the 2010s) and lobster (since the 1970s) harvesting catches have been observed and are primarily attributed to overexploitation (Haimovici et al. 2014a; Cruz et al. 2020; Eggertsen et al. 2024).

The harvesting of other crustaceans and molluscs, for example, the gastropods *Stramonita brasiliensis* (Linnaeus, 1767) and *Turbinella laevigata* Anton, 1838 as well as the crabs *Menippe nodifrons* Stimpson, 1859 and *Carpilus coralinus* (Herbst, 1783), are poorly documented. These species are traditionally collected by local people for their own consumption or are sold at local markets to increase their income (Manzoni and Lacava 1998; Oshiro 1999; Pinheiro et al. 2016; Silva and Martins 2017). A few species (5%) are also used by local people as popular medicine, including the endangered coral *Mussismilia harttii*, gastropods (e.g. *Cassis tuberosa* (Linnaeus, 1758), *Turbinella laevigata*), crabs (e.g. *Plagusia depressa* (Fabricius, 1775)), sea stars (e.g. *Echinaster* spp., *Oreaster reticulatus*) and sea urchins (e.g. *Echinometra lucunter* (Linnaeus, 1758), *Lytechinus variegatus*). The amount and location of this fishing activity is also very scarce (Alves and Dias 2010).

The lack of ecological, biological and fishery data on many harvested invertebrate species is a major threat to these populations because it prevents the evaluation of conservation status of the target species and the regulation of fishing activities. This is even more worrying considering that harvesting of 34% of the species among cnidarians, echinoderms, sponges, ascidians and polychaetes are illegal (Fig. 6.2, Supplementary Material 6.1). Most of these species, plus crustaceans and molluscs legally harvested, go to the aquarium trade (66%) or to the markets of souvenirs, artcraft and decoration (20%) (Dias et al. 2011; Gurjão and Lotufo 2018; Pinheiro et al. 2018). Despite the lack of information on most invertebrate fisheries, it is known that many species are heavily exploited (Alves and Dias 2010; Dias et al. 2011; Ventura et al. 2013), which probably have caused declines in local populations. A noteworthy case is the local extinction of *C. gigantea* in Arraial do Cabo, in Southeastern Brazil, caused by the overexploitation by the marine ornamental aquarium trade (Gasparini et al. 2005). Moreover, the harvesting of foundation species, such as mussels, gorgonians, corals and sponges, causes indirect effects of many associated species because of habitat destruction (Pinheiro et al. 2018).

Although only 27 species of macroalgae are known to be harvested from rocky reefs either for the aquarium trade or the food industry (Fig. 6.1), the intense exploitation depleted populations of red algae from the genera *Gracilaria* and *Hypnea* (Marinho-Soriano 2017). Harvesting of these taxa occurs since the 1960s and is destinated to the extraction of agar and carrageenan (Marinho-Soriano 2017). Other species, including green and brown macroalgae, are collected for the aquarium market with scarce information on this fishery. Nonetheless, the conservation status of all harvested macroalgal species has not been evaluated yet, and harvesting is legal with no regulating norms (Figs. 6.1 and 6.2). The scarcity of data on harvesting and lack of regulating laws is a major threat to macroalgal populations and consequently to the system as a whole, since macroalgae provide shelter for many other organisms and play an important role in biogeochemical processes (Coleman and Williams 2002; Hardison et al. 2011).

Marine Aquaculture of Rocky Reef Species

The overexploitation and intense demand of some benthic organisms for consumption and the food industry has promoted the development of aquaculture (Henriques et al. 2004; Carranza et al. 2009; Marinho-Soriano 2017). Marine aquaculture is responsible for nearly 20% of Brazil's cultivated production (MPA/Brasil 2010). Bivalve farming alone has contributed for 18% of the national production, as of 2018 (IBGE 2023; Valenti et al. 2021).

There are six reef associated species that are commercially cultivated in Brazil, including four species of bivalve and two species of macroalgae (Table 6.1). Bivalve farming, especially of the mussel *Perna perna* (Linnaeus, 1758) and the non-native oyster *Magallana gigas* (Thunberg, 1793), represents the main farming activity among rocky reef species, contributing to 17% to the marine aquaculture production (MPA/Brasil 2010). Farming of mussels, oysters and scallops started in 1990 and is concentrated in the South (SC, PR) and Southeast (SP, RJ and ES) of Brazil (FAO 2007; Rupp and Parsons 2006). Macroalgae aquaculture has initiated in the late 1990s with the development of *Crassiphycus birdiae* (E.Plastino and E.C.Oliveira) Gurgel, J.N.Norris and Fredericq, 2018 farming in the Northeast (CE, RN and PB), and of the non-native *Kappaphycus alvarezii* (Doty) Doty ex P.C.Silva, 1996 in the

Table 6.1 List of commercial and artisanal rocky reef farmed species in the Brazilian coast with known exploitation uses

Taxonomic group	Species name	Use	Distribution
Invertebrates			
Bivalve	*Crassostrea rhizophorae* (Guilding, 1828)	Consumption	Native to Brazil
	Magallana gigas (Thunberg, 1793)	Consumption	Non-native
	Nodipecten nodosus (Linnaeus, 1758)	Consumption	Native to Brazil
	Perna perna (Linnaeus, 1758)	Consumption	Non-native
Macroalgae			
Red algae	*Kappaphycus alvarezii* (Doty) Doty ex P.C.Silva, 1996	Food industry	Non-native
	Crassiphycus birdiae (E.Plastino & E.C.Oliveira) Gurgel, J.N.Norris & Fredericq, 2018	Food industry	Native to Brazil

Species are listed as the current accepted name according to the World Register of Marine Species (last access: November 2020)

Southeast (SP and RJ) and South (SC) of Brazil (Hayashi and Reis 2012; Marinho-Soriano 2017).

Farming of the mussel *Perna perna* used to rely on collecting the seeds (i.e. juveniles) directly from natural mussel beds. This has caused a dramatic decline in the living stocks, although this species is currently classified as non-native to Brazil (FAO 2007; Bumbeer and Rocha 2016; Oricchio et al. 2019). Most of the mussel seeds are currently obtained using seed collectors deployed in the water column, minimizing the impact on natural mussel beds (FAO 2007). Despite the efforts to minimize environmental impacts of mussel farming, such activity can increase fish abundances in its surroundings (Kuhnen et al. 2019). Moreover, the high deposition of organic matter by bivalve faeces production has altered the structure of sediment meiofauna by decreasing species richness and diversity, but no effect was detected on macrofauna (da Costa and Nalesso 2006, Netto and Valgas 2010).

Bioinvasion is another environmental concern associated to mussel and oyster farming. The substrate of the structures in the farms (i.e. shells and artificial material) provides novel habitats for fouling organisms, facilitating the settlement of non-native species, such as tunicates (Rocha et al. 2009). Furthermore, the non-indigenous oyster *Magallana gigas* have spread onto natural rocky reefs, probably because of the oyster farms (Darrigran et al. 2020; Melo et al. 2010). The environmental impact of the oyster on natural shores is still unknown but has the potential to outcompete native species and change the structure of local benthic assemblages (Darrigran et al. 2020).

Regarding macroalgal farming, there is no evidence of negative environmental impacts either of the native species or the non-native *Kappaphycus alvarezii* (Castelar et al. 2009a, b; Silva et al. 2010). However, there has been little effort to monitor and evaluate potential impacts of such activity (Silva et al. 2010; Pellizzari and Reis 2011; Simioni et al. 2019). On the other hand, macroalgal farming has the potential to improve water quality by removing excess nutrients from areas contaminated by nitrogen and phosphorus (Buschmann et al. 2008; Alencar 2010; Monteiro 2011).

Coastal Pollution

Contamination of coastal systems is an increasing issue worldwide. In Brazil, pollution has historically been associated mostly to industrial effluents, untreated domestic waste and agriculture (Martinez et al. 2022). The increased population on the coast and removal of natural green areas on land has led to an increase in several contaminants within estuarine and coastal areas (e.g. Lamparelli et al. 2001; Moles and Bunge 2002; Buruaem et al. 2012; Pereira et al. 2016; Rosa Filho et al. 2018). These contaminants include nutrients (e.g. nitrate, ammonium, phosphate), organic matter (e.g. POM, SOM, TOC), biological contaminants (e.g. faecal coliforms, heterotrophic bacteria, yeast), metals (e.g. Cu, Fe, Hg), hydrocarbons (e.g. PAHs, PACs, oils), organometallics (e.g. TBT, TPT, DBT), pesticides (e.g. DDT, HCB, Aldrin), industrial contaminants (PCBs, Surfactants, PFCs), marine debris (e.g. plastic bags, nylon, microplastic) and cosmetics and pharmaceutics (e.g. sunscreen, ibuprofen, acetaminophen). Research on Brazilian rocky shores has predominantly concentrated on pollution from metals, hydrocarbons and nutrients (Fig. 6.3).

Although coastal contamination in Brazil has a long-drawn-out history, data regarding its impact on organisms inhabiting rocky shores remains scarce. The current knowledge of pollution is based on responses measured in a few taxonomic groups at both the species (fauna: 9 classes among 5 phyla, $N = 89$ responses; flora: 3 classes among 3 algal phyla, $N = 141$ responses) and assemblage level (fauna: 8 assemblage groups, $N = 36$ responses; flora: 4 assemblage groups, $N = 12$ responses) (Figs. 6.4, 6.5, 6.6, and 6.7).

Contamination Impacts on Benthic Invertebrates and Fish

Most of the available data on the impacts of pollutants on faunal organisms have derived from studies assessing the responses in molluscs, crustaceans and echinoderms (Figs. 6.4 and 6.5). In studies with molluscs, especially bivalves, it has been predominantly observed that contamination does not significantly affect oxidative stress, which was the response most frequently quantified (Fig. 6.4). Also, contamination by nutrients and faecal coliforms seems to have no effect on cardiac activity or cytotoxicity on mussels (Abessa et al. 2005). On the other hand, contamination by metals or other contaminants associated with urbanization may, in some cases, increase oxidative stress, cytotoxicity and filtration and biodeposition rates in these organisms (Rola et al. 2012; Martinez et al. 2019). Furthermore, metals can cause

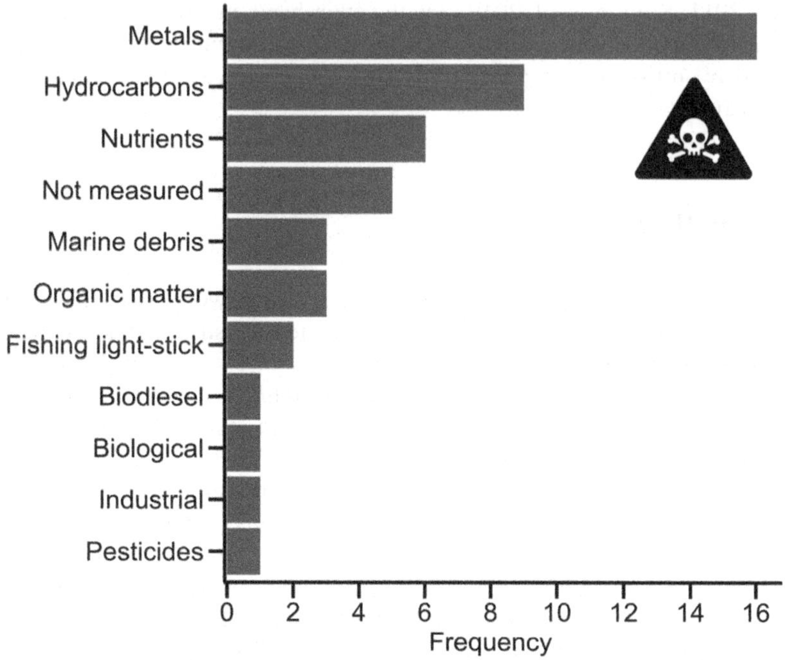

Fig. 6.3 Frequencies of the type of contaminant measured in Brazilian rocky shore studies. Data was extracted from 41 scientific publications ranging from 2004 to 2022 that measured organisms' responses to contamination either through mensurative observational or manipulative experiments (Supplementary Material 6.2). The graph used in this figure was generated with R version 4.3.2 and subsequently edited in Inkscape version 1.3.2. (Icons were sourced from SVG Repo and the IAN Symbol Libraries)

declines in mussel growth and increase in developmental abnormality (Szalaj et al. 2017). Other contaminants, like hydrocarbons, have shown to increase genotoxicity in these bivalves (Francioni et al. 2007). Responses measured in gastropods have shown an increase in reproductive disorder (Fig. 6.4), associated with contamination by organometallics present in antifouling paint (Castro et al. 2012). Overall, contamination associated with heavily urbanized areas (hydrocarbons, PCBs, metals and DDTs) causes morphological alterations in limpet shells, decreasing shell height, thickness and weight, while increasing shell specific gravity (Begliomini et al. 2017). Such contamination may also increase oxidative stress and genotoxicity in these animals (Begliomini et al. 2017). Periwinkle populations and gastropod assemblages, however, appear to be resilient to hydrocarbon contamination (Figs. 6.4 and 6.5). Following the significant oil spill incident in the Northeast of Brazil in 2019, periwinkles have colonized rocky substrates in the affected areas (Gusmao et al. 2021). Likewise, diesel oil did not affect colonization of local gastropod assemblage (Grande et al. 2012).

Evidence from research indicates that coastal contamination affects various crustacean groups (Figs. 6.4 and 6.5). Contamination by hydrocarbons (crude oil)

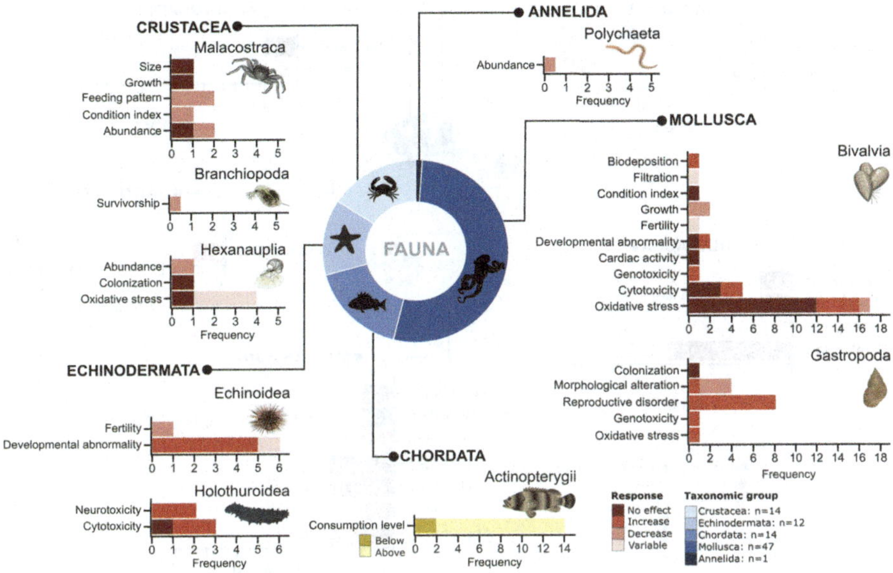

Fig. 6.4 Frequency of the response endpoints measured in benthic invertebrates and vertebrates associated with rocky shores from Brazilian studies that investigated the effects of contamination on the organisms at the individual or population level. Data was extracted from 41 scientific publications ranging from 2004 to 2022 that measured organisms' responses to contamination either through mensurative observational or manipulative experiments (Supplementary Material 6.2)

decreased abundances of female crabs and barnacles but did not seem to affect abundances of male crabs and crabs' size (Santana et al. 2022). Colonization by barnacle populations (Fig. 6.4), as well as by assemblages of decapods and tanaidaceans (Fig. 6.5), appears to be resistant to hydrocarbon contamination (Grande et al. 2012; Santana et al. 2022). However, hydrocarbon exposure (diesel oil) may impact colonization by assemblages of isopods and, in some cases, of amphipods (Fig. 6.5) (Grande et al. 2012). Furthermore, crabs (Decapoda; Fig. 6.4) showed a decline in body condition index and changes in feeding patterns (decrease in gut fullness and food diversity) with an increase in microplastic contamination (Barros and dos Santos Calado 2020; Barros et al. 2020).

Pollution research on echinoderms (Fig. 6.4) has revealed that exposure to contaminants such as fishing light-sticks, biodiesel, and microplastics results in increased developmental abnormalities and reduced fertility in sea urchins (Cesar-Ribeiro and Palanch-Hans 2010; Leite et al. 2011; Nobre et al. 2015; Cesar-Ribeiro et al. 2017). Additionally, exposure to metal contaminants has been associated with heightened neurotoxicity and cytotoxicity in sea cucumbers (Pereira et al. 2017).

Available studies on the effects of hydrocarbons on polychaetes indicate that crude oil leads to reductions in population abundance (Fig. 6.4) and assemblage diversity (Fig. 6.5) (Grande et al. 2012; Craveiro et al. 2021). In contrast, diesel oil

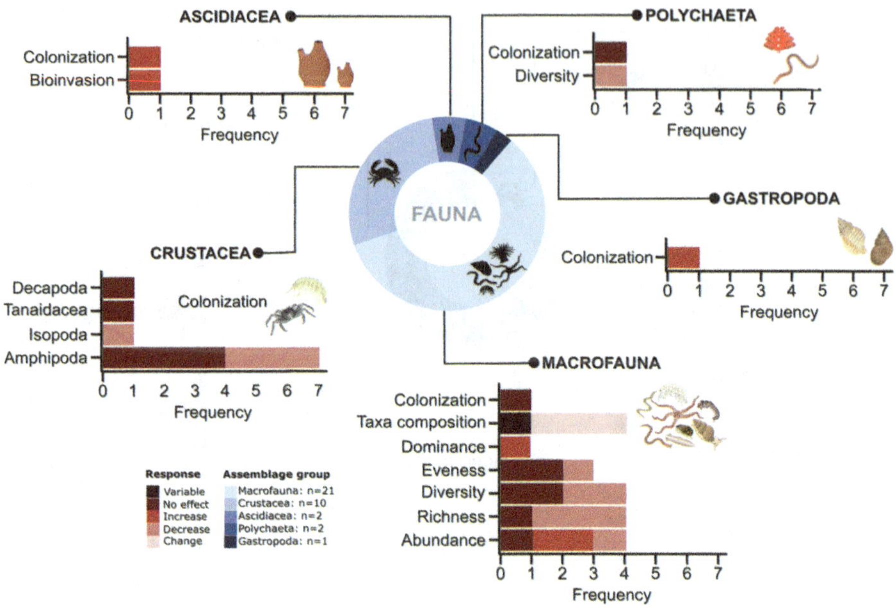

Fig. 6.5 Frequency of the response endpoints measured in benthic invertebrates associated with rocky shores from Brazilian studies that investigated the effects of contamination on the organisms at the assemblage level. Note that "colonization" was the only response endpoint assessed in crustaceans and reported for each crustacean assemblage. Data was extracted from 41 scientific publications ranging from 2004 to 2022 that measured organisms' responses to contamination either through mensurative observational or manipulative experiments (Supplementary Material 6.2). The graphs used in this figure were generated with R version 4.3.2 and subsequently edited and assembled in Inkscape version 1.3.2. (Icons were sourced from SVG Repo and the IAN Symbol Libraries)

exposure does not appear to influence the colonization of these assemblages (Lira et al. 2021).

In the case of chordate organisms, studies have demonstrated an increase in colonization and bioinvasion of ascidian assemblages in environments enriched with nutrients and organic matter (Fig. 6.5) (Marins et al. 2010). In another context, most rocky reef fishes sampled within or near urban areas exhibited metal concentrations exceeding the safety thresholds established for human consumption (Fig. 6.4) (Medeiros et al. 2012; de Santana et al. 2017).

Research on macrofaunal assemblages of rocky shores has reported adverse impacts on structural patterns (Fig. 6.5). Crude oil exposure has been associated with increased macrofaunal abundances, while nutrient pollution tends to diminish their numbers (Zalmon et al. 2011; Craveiro et al. 2021). Furthermore, nutrient enrichment can lead to reduced diversity, increased dominance and changes in taxa composition (Zalmon et al. 2011). Exposure to crude oil has also been documented to alter macrofaunal assemblage composition, along with causing a reduction in both diversity and evenness (Craveiro et al. 2021). The latter two impacts, however, were observed exclusively in the macrofauna associated with one of the two algal species studied (Craveiro et al. 2021).

Contamination Impacts on Benthic Algae

Research into pollution effects on rocky macroalgae has largely focused on assessing photosynthetic activity and photosynthetic components, yielding inconsistent results (Fig. 6.6). In nearly half of the observed cases, contamination did not impact photosynthetic activity. Where changes were noted, the results varied over time or contamination treatments (e.g. Schmidt et al. 2015; Costa et al. 2017b), with some cases showing decreased or, less frequently, increased photosynthetic activity in response to contamination (e.g. Scherner et al. 2012; Santos et al. 2015). Conversely, photosynthetic components of Florideophyceae algae were, in most cases, reduced with increased exposure to contaminants (Bouzon et al. 2012; Gouveia et al. 2013; Santos et al. 2015). Ulvophyceae and Phaeophyceae algae showed either an increase or variable responses to contamination (Felix et al. 2014; Costa et al. 2017a, b; Pilatti et al. 2016).

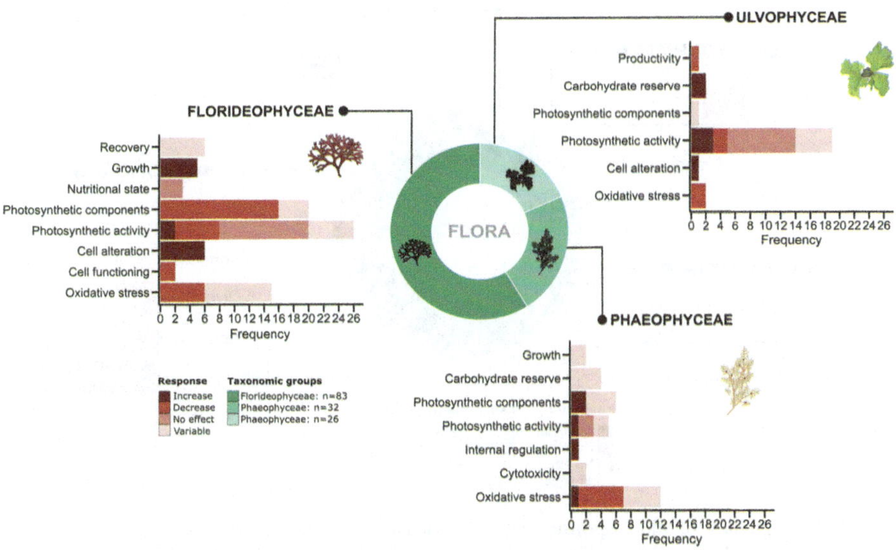

Fig. 6.6 Frequency of the response endpoints measured in macroalgae associated with rocky shores from Brazilian studies that investigated the effects of contamination on the organisms at the individual or population level. Data was extracted from 41 scientific publications ranging from 2004 to 2022 that measured organisms' responses to contamination either through mensurative observational or manipulative experiments (Supplementary Material 6.2). The graphs used in this figure were generated with R version 4.3.2 and subsequently edited and assembled in Inkscape version 1.3.2. (Icons were sourced from SVG Repo and the IAN Symbol Libraries)

Oxidative stress, measured by carotenoid, polyphenol, flavonoid concentrations, phenolic or flavonoid compounds, often decreased with exposure to metal or hydrocarbon, indicating elevated basal stress (Santos et al. 2015; Pilatti et al. 2016; Costa et al. 2017a). In green Ulvophyceae algae, other alterations included increased cell alteration and carbohydrate reserve and reduction in productivity with elevated metal or hydrocarbon contamination (Andrade et al. 2004; Pilatti et al. 2016). Growth, cytotoxicity and carbohydrate reserve responses to metal exposure were variable in brown Phaeophyceae algae (Andrade et al. 2004; Costa et al. 2017a, b). In red Florideophyceae algae, metal contamination has been linked to an increase in cellular alterations, with frequent observations of diminished growth and impaired cellular functioning (Bouzon et al. 2012; Gouveia et al. 2013; Felix et al. 2014; Santos et al. 2015).

Studies on the effects of nutrient enrichment and urban contamination have documented a decline in macroalgal species richness and diversity (Fig. 6.7) (Martins et al. 2012). These conditions may cause an increase in the abundance of foliose and siphonous algae while leading to a reduction in canopy-forming algal assemblages (Vasconcelos et al. 2019). Specifically, nutrient enrichment has been found to elevate the abundances of green algae (Chlorophyta) and diminish the abundances of brown algae (Ochrophyta) (Vasconcelos et al. 2019).

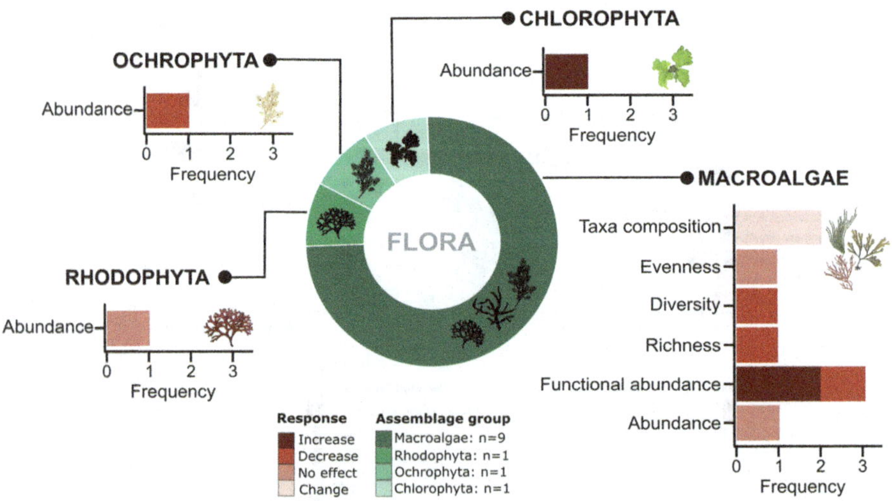

Fig. 6.7 Frequency of the response endpoints measured in macroalgae associated with rocky shores from Brazilian studies that investigated the effects of contamination on the organisms at the assemblage level. Data was extracted from 41 scientific publications ranging from 2004 to 2022 that measured organisms' responses to contamination either through mensurative observational or manipulative experiments (Supplementary Material 6.2). The graphs used in this figure were generated with R version 4.3.2, and subsequently edited and assembled in Inkscape version 1.3.2. (Icons were sourced from SVG Repo and the IAN Symbol Libraries)

Spills

Spills of crude oil and its derivatives are an inherent risk in maritime shipping. The increase in the flux of vessels in the ocean, seas and coastal regions has, consequently, led to an increase in the occurrence of these maritime incidents (Zanella 2010). Given the extensive exploration, production and transport of oil and its byproducts, numerous spills have led to the pollution of Brazil's marine, coastal and oceanic areas, causing harm to marine and estuarine organisms, as well as affecting fishing, mariculture and tourism sectors. The initial records of oil spill incidents in Brazil date back to the 1970s (Fig. 6.8). The earliest events, primarily related to maritime transportation, accounted for the most substantial volumes of oil released. In subsequent years, incidents have occurred in pipelines, terminals and refineries across various regions of the country, although the volumes involved have been comparatively smaller (Fig. 6.8, Supplementary Material 6.3). One of the major oil spills in Brazil occurred recently, in 2019, reaching 1009 sites, in more than 130 municipalities, in 11 states, totalling an area of more than 3000 km of the Brazilian Coast (IBAMA 2020a).

While incidents involving oil and its products have occurred, quantifiable impacts on rocky shore organisms remain limited. Research focusing on the effects of hydrocarbons on polychaetes suggests that crude oil exposure results in a decline in population numbers and diversity (Grande et al. 2012; Craveiro et al. 2021). Interestingly, overall macrofaunal populations have been noted to increase following crude oil exposure (Craveiro et al. 2021). However, this exposure also caused shifts in macrofaunal community structure and lead to decreased diversity and evenness (Craveiro et al. 2021). Moreover, barnacle and lobster larvae, as well as female crabs, exhibit considerable sensitivity to oil, with molluscs such as mussels (*Mytilus*)

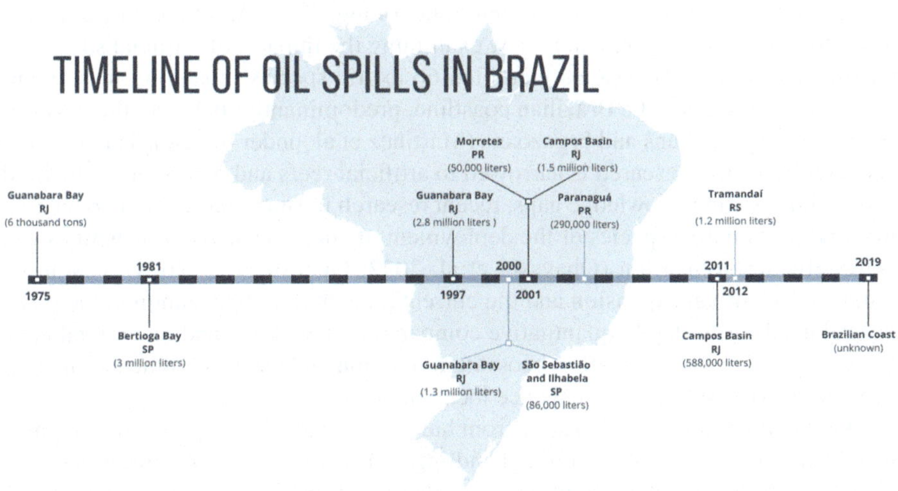

Fig. 6.8 Timeline of accidents with oil and oil products in Brazil

and oysters (*Crassostrea*) experienced reduced reproductive rates (Baussant et al. 2011; Vignier et al. 2017; Garcia et al. 2020; Santana et al. 2022).

In November 2015, a catastrophic environmental disaster unfolded with the collapse of the Fundão dam, which led to the release of around 40 million cubic metres of iron mining waste into the Rio Doce Basin (Hatje et al. 2017). The contaminated water and sediments reached the ocean, but the impacts on rocky shore organisms remain unknown.

Habitat Modification

The modification of rocky shore habitats in Brazil due to human activities such as construction and tourism is a critical environmental issue with significant ecological consequences. The global trend towards urban expansion brings about the construction of artificial structures that can disrupt natural ecosystems, leading to biodiversity loss, changes in habitat connectivity and facilitating bioinvasion (Chapman 2003; Airoldi et al. 2015; Gittman et al. 2016; Bishop et al. 2017). In Brazil, the rapid expansion of artificial structures, including port developments and various marine constructions, poses a significant threat to the biodiversity of local rocky shores. Along the São Paulo coastline, 244 km is now occupied by artificial constructs such as breakwalls, jetties, wharves and aquaculture apparatus (Pardal et al. 2024). These modifications could be inducing significant ecological shifts, yet the comprehensive effects on the biota of rocky shores remain poorly understood.

The construction of structures over rocky shores, for instance, casts shadow on the biodiversity of these ecosystems. Shading may lead to reduced macroalgal biomass and alter community structure, favouring filter feeders over grazers (Pardal-Souza et al. 2017). Such structural changes suggest a potential loss in ecosystem complexity and alterations in ecosystem functioning. These AS also create opportunities for bioinvasion. A recent review examining the impacts of artificial structures on Brazil's coastal ecosystems identified 66 exotic species inhabiting these man-made substrates along the Brazilian coastline, predominantly rocky benthic invertebrates such as ascidians and bryozoans (Martinez et al. under review). This revision also highlights that research is narrowed to artificial reefs and fish ecology. Beyond addressing existing knowledge gaps, recent research in Brazil has demonstrated that the seeding of native species or the deployment of shells on artificial structures can help mitigate bioinvasion (Ohayashi et al. 2022; Queiroz et al. 2024). Given the rapid pace of urban expansion and the current gaps in our understanding, there is a critical need for in-depth, quantitative comparison of artificial and natural habitats, focusing on variables that reflect ecosystem function and the development of marine eco-engineering solutions to enhance local biodiversity.

Habitat alteration may also arise from human recreational activities, where physical disturbances such as trampling, handling and removal of organisms, along with alterations to local water conditions (e.g. increased turbulence and water motion), can impact rocky shore organisms. Trampling has been observed to reduce chlorophyll-a fluorescence and the concentration of photosynthetic pigments in *Sargassum* beds (Barradas et al. 2018). While these effects varied by location, they could suggest that trampling affects the health of these algal assemblages (Barradas et al. 2018). Such disturbance does not seem, however, to affect the abundance and distribution of biofilms, probably due to their inherent resilience and the substrate's roughness (Leite et al. 2012).

This physical impact extends to intertidal fauna. Even low-intensity trampling has been demonstrated to decrease abundances of the barnacle *Chthamalus bisinuatus* (Pilsbry, 1916) (Ferreira and Rosso 2009). Notably, the abundances of this barnacle did not return to natural levels even 7 months after trampling simulations ceased (Ferreira and Rosso 2009). In contrast, trampling on beds of the invasive bivalve *Isognomon bicolor* (C. B. Adams, 1845) did not impact their populations and was observed to have a positive effect on the species richness of the associated macrofauna (Ferreira and Rosso 2009). Despite the observed impacts of recreational activities, environmental education has proven effective at reducing the impact of visitors on rocky shores, as demonstrated in areas where such programmes are implemented, leading to more responsible visitor behaviour and less environmental damage (Silva and Ghilardi-Lopes 2012).

Biological Invasion

Currently, biological invasions have been considered one of the major threats to marine biodiversity (Ruiz and Carlton 2003). Due to its abundance and geographical dispersion, invasive species may cause several impacts in invaded coastal systems, including changes to native species' growth, reproduction and survival and to community abundance and diversity (Grosholz 2002).

In the past few decades, there was an increase in species transfer through ballast water or biofouling on ship hulls (Ruiz and Carlton 2003; Carlton and Ruiz 2005). This occurred mainly due to the growing shipping traffic because of the global trade expansion, which is also true for Brazil (Teixeira and Creed 2020). Moreover, other human-mediated activities can also act as vectors of species introduction and dispersion, such as aquaculture, the movement of dry docks, construction of artificial structures in the coastline, exploratory drilling platforms and the aquarium trade (Williams et al. 2013; Airoldi et al. 2015).

Non-indigenous species (NIS) can be classified into four categories based on their invasive potential, according to the classification system adopted by the Ministry of the Environment of Brazil (MMA), as outlined by Lopes and Villac (2009): *Contained*—NIS that has been detected only in controlled artificial environments, totally or partially isolated from the natural environment (e.g. commercial aquarium, cultivation for scientific purposes, vessel ballast water tanks); *Detected*—NIS detected in the natural environment, however, without any subsequent increase in its abundance and/or dispersion, or without subsequent information on the species population status (isolated record); *Established*—NIS detected on a recurrent basis, with a complete life cycle in the wild and evidence of population increase over time in a restricted or wide region, but without apparent ecological or socio-economic impacts; *Invasive*—NIS that interferes with the survival capacity of other species in a wide geographic region or even in a specific area, due to its abundance or geographical dispersion (Elliott 2003), or when the established species causes measurable impacts on socioeconomic activities or on human health.

In Brazil, there are a total of 138 marine NIS inhabiting or associated with hard bottom substrate on the Brazilian Coast (Fig. 6.9, Supplementary Material 6.4). The majority of NIS are benthic sessile species (118 spp.), with a few planktonic (12 spp.) and nektonic ones (8 spp.). Most species are those with recurring or spatial expansion records on the coast (63 established species), or records are punctual with no evidence of spatial expansion (40 detected species). A quarter of the NIS (i.e. 35 species among ten classes) are currently classified as invasive. Most invasive species are ascidians (11 spp.) and cnidarians (8 spp.), followed by molluscs (4 spp.) and arthropods (4 spp.) (Fig. 6.9).

Building on the recent inventory of non-indigenous species (NIS) in Brazil by Teixeira and Creed (2020), five additional species have been detected in Brazilian waters, comprising one sea sponge, two cnidarians, one bivalve and one ascidian. Three of these species—*Heteropia glomerosa* (Bowerbank, 1873) detected in 2005 (Klautau et al. 2020), *Perna viridis* (Linnaeus, 1758) detected in 2018 (Messano

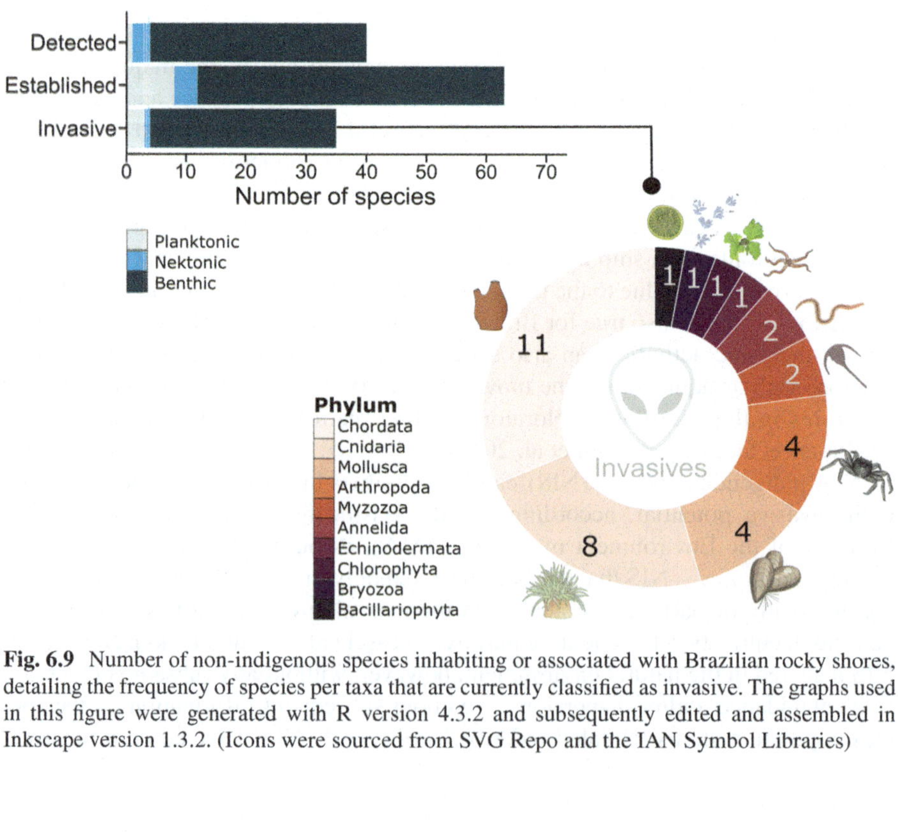

Fig. 6.9 Number of non-indigenous species inhabiting or associated with Brazilian rocky shores, detailing the frequency of species per taxa that are currently classified as invasive. The graphs used in this figure were generated with R version 4.3.2 and subsequently edited and assembled in Inkscape version 1.3.2. (Icons were sourced from SVG Repo and the IAN Symbol Libraries)

et al. 2019), and *Didemnum* sp. RJAC detected in 2016 (Granthom-Costa et al. 2023)—were reported to the coast of Rio de Janeiro State. The other two species are the octocorals *Briareum hamrum* (Gohar, 1948) and *Sarcothelia* sp. n.i. found on rocky reefs in the Bay of Todos os Santos, Bahia, in 2018 (Menezes et al. 2021). The calcareous sponges *Paraleucilla magna* Klautau, Monteiro and Borojevic, 2004 and *Sycettusa hastifera* (Row, 1909), which were considered cryptogenic (Klautau et al. 2004; Teixeira and Creed 2020), have recently been reclassified as NIS (Klautau et al. 2020). Additionally, the bryozoans *Schizoporella errata* (Waters, 1878), *Buskia socialis* Hincks, 1886, *Conopeum reticulum* (Linnaeus, 1767) and *Scruparia ambigua* (d'Orbigny, 1841) were recently classified NIS in the categories invasive, detected, established and detected, respectively (Xavier et al. 2021).

Cases of Invasive Species on Brazilian Rocky Shores

Here we have compiled knowledge on a few invasive species on Brazilian rocky shores that have shown notable impacts on local native biota or on economic activities, which have demonstrated rapid spread along in Brazil or around the world.

The Ascidian *Styella plicata* (Lesueur, 1823)

The ascidian *Styela plicata*, probably native to the Indo-Pacific, is widely distributed around the world (Fig. 6.10). This species is commonly found in shallow, sheltered waters and is particularly abundant in port areas (Monniot and Monniot 1997; Carlton 2009; Pineda et al. 2011). The first record of this species in Brazil was in the nineteenth century, in Rio de Janeiro (Rocha and Kremer 2005). Currently, there are records of this species in the states of Santa Catarina, Paraná, São Paulo, Rio de Janeiro and Bahia (Rocha and Kremer 2005; Barros et al. 2009). On mussel and oyster farms, where it is very common, *S. plicata* is considered to be a major nuisance because it occurs in large numbers, competing with mussels and oysters and increasing the cost for maintenance (Rocha et al. 2009).

Sun Corals *Tubastraea* spp.

The sun coral—scleractinian corals of the genus *Tubastraea* (Fig. 6.11)—is one of the most significant cases of marine bioinvasion in Brazil. *Tubastraea* spp. were first detected in southeastern Brazil in the 1980s on oil platforms (Castro and Pires 2001). The species *Tubastraea tagusensis* Wells, 1982 and *Tubastraea coccinea* Lesson, 1830 are both native to the South Pacific, from Galapagos Islands (Cairns 2000) and the Fiji Archipelago (Cairns 1994), respectively. Today, *T. coccinea* is considered cosmopolitan, as the species can be found in tropical regions all over the world.

Fig. 6.10 The solitary ascidian *Styela plicata* on artificial structures and in experimental panels on Guanabara Bay, Rio de Janeiro, Brazil. (Photos: Luciana Altvater and Monalisa Sousa Pinto de Oliveira)

Fig. 6.11 Sun corals of the genus *Tubastraea* on rocky shores and artificial structures in Arraial do Cabo, Rio de Janeiro, Brazil. (Photos: Luciana Altvater)

Over the past 40 years, sun corals have spread discontinuously along the Brazilian Coast (Coelho et al. 2022), colonizing both natural and artificial substrates. Currently, these species have been registered in Santa Catarina, São Paulo, Rio de Janeiro, Espírito Santo, Bahia, Sergipe, Pernambuco, Rio Grande do Norte e Ceará (de Paula and Creed 2004; Da Silva et al. 2014; Miranda et al. 2016; Soares et al. 2016; Capel et al. 2019; de Oliveira Soares et al. 2020; Teixeira and Creed 2020). Moreover, a recent study has confirmed the presence of a third species of sun coral in Brazil, previously misidentified as *Tubastraea tagusensis*, occurring in Arraial do Cabo, Rio de Janeiro (Bastos et al. 2022).

Some documented impacts of sun coral on native biota include competition for space, dislodgement of native species and changes in local biodiversity (Lages et al. 2011; Mantelatto et al. 2011; Silva et al. 2019; Tanasovici et al. 2020). Additionally, direct contact with sun coral colonies has been shown to cause necrosis in native corals (Creed 2006; Miranda et al. 2016). Given their invasive potential, *Tubastraea* spp. have been identified as a priority for control measures in Brazil. In 2018, the Ministry of the Environment (MMA), in collaboration with other governmental environmental agencies including IBAMA and ICMBIO, formulated a National Plan for the Prevention, Control, and Monitoring of Sun Coral (IBAMA 2020b).

The Green Mussel *Perna viridis* (Linnaeus, 1758)

The Asian green mussel *Perna viridis*, native from India to Southeast Asia, is a known invader in several countries. Recently, some specimens were found on artificial structures in Guanabara Bay, Rio de Janeiro (Fig. 6.12) (Messano et al. 2019). In 2022, *P. viridis* was also found at a mussel farm in Arraial do Cabo, Rio de Janeiro,

Fig. 6.12 The green mussel *Perna viridis* on artificial substrata within Guanabara Bay, Rio de Janeiro, Brazil. (Photos: Luciana Altvater)

approximately 150 km from its initial point of introduction (Santos et al. 2023). Although the impacts of this species have not yet been assessed in Brazil, it has been recognized as a high-risk invader in various global locations (Dias et al. 2018).

The Winged Oyster *Isognomon bicolor* (C. B. Adams, 1845)

Isognomon bicolor, a small bivalve native to the Western Central Atlantic Coast, was first recorded along the Brazilian coast in 2002 (Domaneschi and Martins 2002). This bivalve, characterized by rapid population growth and continuous reproduction, is found in high abundances on Brazilian intertidal rocky reefs on bare rock and within rocky pools or biogenic microhabitats such as calcareous algae and empty shells of barnacles (Breves-Ramos et al. 2010; Martinez 2012; Queiroz et al. 2022). It has spread its distribution range from the northeast coast (Rio Grande do Norte) to the southern coast (Santa Catarina) (Domaneschi and Martins 2002; Breves-Ramos et al. 2010; Loebmann et al. 2010; Martinez 2012; Lima and Passos 2021; Silva e Cavalcante and Gomes 2022; Queiroz et al. 2023). Additionally, it has been recorded on artificial buoys along the coast of Rio Grande do Sul (Agostini and Ozorio 2016).

By colonizing the rocky space in large densities, the winged oyster has contributed to the decline of native species and has altered the structure of the benthic community (Lopez et al. 2014). Furthermore, there is evidence that *I. bicolor* alters local predator-prey dynamics, potentially leading to significant changes in the structure of local benthic assemblages. Notably, predatory gastropods such as *Stramonita haemastoma* (Linnaeus, 1767) and *Trachypollia nodulosa* (C. B. Adams, 1845) prey on the invasive bivalve, albeit at varying intensities (Lopez et al. 2010). This predation could consequently alleviate the pressure on native species and alter local species dynamics and community structure.

The Fan Worm *Branchiomma luctuosum* (Grube, 1870)

The *Branchiomma luctuosum*, native from the Red Sea, was first recorded in Brazil in 2003 (Fig. 6.13), in São Paulo (Nogueira et al. 2006). After this, the species was registered in Rio de Janeiro and in Paraíba states, Northeast Brazil (Costa-Paiva 2006; Assis et al. 2012). This species can dominate fouling communities on estuarine shallow waters (Oricchio et al. 2019), monopolizing the substrate and thereby impeding colonization by native species (Corrêa et al. 2023).

The Green Macroalgae *Caulerpa scalpelliformes* (R.Brown ex Turner) C.Agardh, 1817

The tropical macroalga *Caulerpa scalpelliformes* was a domestic introduction. Its native distribution extends from Piauí to Espírito Santo. However, in 2001, the species was found in a site at Ilha Grande Bay, Rio de Janeiro, approximately 400 km

Fig. 6.13 The tubiculous fan worm *Branchiomma luctuosum* on experimental panels on Guanabara Bay, Rio de Janeiro, Brazil. (Photo: Denny Fardin)

from its southern limit. In the new area, *C. scalpelliformes* spread rapidly on the rocky shore, altered the macroalgae community structure, by dislodging and replacing species such as the dominant *Sargassum vulgare* C. Agardh, 1820 (Falcão and Széchy 2005; Vasconcelos et al. 2011).

Climate Change

Climate change may impact organisms on rocky shores by modifying sea and rock surface temperatures, wave dynamics, salinity, pH and inundation levels. Research on the effects of climate change on Brazilian rocky shores remains scarce (Coutinho et al. 2016). However, recent studies and existing data on environmental shifts and their influence on coastal rocky organisms provide a basis to infer potential future biotic changes due to climate change.

Notably, there has been a documented decrease in the diversity of recruiting species correlated with increased sea surface temperature (SST) and a decline in recruitment abundances during cold front events (Mazzuco et al. 2015, 2019). Similarly, the structure of macrofauna assemblages exhibits variation with SST and wave fetch along the Southeast coast, showing effects at both local (metres) and regional (kilometres) scales (Pardal et al. 2023).

Among climate environmental pressures, species-specific responses also emerge. Climate change has intensified western boundary currents in Brazil (Wu et al. 2012; Baumann and Doherty 2013), potentially facilitating the southward expansion of tropical fish species such as surgeonfish, damselfish and parrotfish (Luiz-Júnior et al. 2004; Barneche et al. 2009). This shift is likely altering community structures in subtropical rocky reefs through increased grazing (Vergés et al. 2014). Meanwhile, alarming trends have been observed in the coverage of *Sargassum* spp., which has experienced an average annual decline of 2.6%, cumulating in average 52% loss (ranging between 20% and 89%) from 1983 to 2017 (Gorman et al. 2020). The most significant canopy losses were recorded at locations experiencing the highest degrees of coastal warming, where long-term SST increases ranged from 0.17 to 0.5 °C (Gorman et al. 2020). Furthermore, climate projections for southeast and east Brazil up to the year 2100 predict decreases in salinity and pH, coupled with more intense temperature fluctuations (Durante et al. 2018). Such environmental shifts are expected to result in a diminished habitat suitability for the zoanthid *Palythoa caribaeorum* Duchassaing & Michelotti, 1860, leading to a potential contraction of its distribution, with survival likely confined to the Northeast and North regions of Brazil (Durante et al. 2018).

Organisms like the barnacle *Chthamalus bisinuatus* show resilience to rising rock temperatures, since post-settlement mortality and larval physiological condition were not affected under manipulated increased rock temperatures (av. 3 °C above natural conditions) (Leal et al. 2020). Additionally, this species has demonstrated enhanced recruitment rates in response to rising sea surface temperatures (Mazzuco et al. 2015). In contrast, many other tropical rocky shore species (fish, crustaceans, gastropods) exhibited lower physiological performance, limited acclimation capacity and high vulnerability to temperature fluctuations, hinting at potential biotic changes sooner in these regions compared to temperate zones (Vinagre et al. 2016, 2019; Madeira et al. 2018b). Notably, tide pools can be considered ecological traps for many of their inhabiting species since tide pools can reach temperatures higher than the upper thermal limits of most of these species (Vinagre et al. 2016, 2018, 2019). Thus, heat waves and global warming pose a critical risk for many intertidal species that might encounter temperatures surpassing their thermal limits.

Sub-lethal effects of warming, such as heat shock response and biochemical stress markers, have also been documented in the gastropod *Stramonita haemastoma*, suggesting that long-term exposure to elevated temperatures may lead to smaller body sizes and compromised physiological function (Madeira et al. 2018a). Additionally, the invasive shrimp *Lysmata lipkei* Okuno and Fiedler, 2010 displays a high thermal tolerance and acclimation capacity that may pose an advantage over other local native species with increased temperatures (Madeira et al. 2019). This highlights the importance of understanding physiological tolerances under changing climatic conditions, but also the potential consequences for alterations in population dynamics and species interactions.

A study investigating the direct effects of increased temperature and ocean acidification on the coralline alga *Jania rubens* (Linnaeus) J.V. Lamouroux, 1816 and the sea urchin herbivore *Echinometra lucunter* showed that while algal physiology remained mostly stable, the sea urchins experienced elevated metabolic and consumption rates under acidified conditions (Rich et al. 2018). Interestingly, these findings suggest that short-term climate stressors may have a more pronounced impact on herbivore behaviour than on algal palatability (Rich et al. 2018).

Complementing this, predator-prey dynamics, exemplified by a whelk–barnacle system, were found to be influenced by local wave exposure and prey density (Pardal et al. 2022). The whelks' consumption rates decreased with greater wave exposure, particularly at lower prey densities, highlighting that environmental harshness and prey availability are crucial in modulating predatory interactions (Pardal et al. 2022). These results suggest that changes in species interactions associated with climate change may vary according to species-specific traits and spatial scale.

The findings from recent research in Brazil illustrate the urgent need for continued monitoring and research to understand the full scope of climate change impacts on coastal biota, which are essential for informing conservation strategies and managing the resilience of Brazilian rocky shores.

Concluding Remark

Research on human impacts along the Brazilian coast has uncovered substantial and diverse effects on rocky shore ecosystems. However, it is important to highlight that much of the research is spatially fragmented, revealing major local knowledge gaps that must be addressed to better guide sustainable coastal management in Brazil. Future research should broaden to encompass diverse taxonomic groups and incorporate both structural and functional response assessments of the biota and ecosystems. Furthermore, there is a need to understand the interactive effects of multiple stressors and implement long-term monitoring research aiming at adaptive management of human impacts. Such comprehensive research is essential to inform policy and marine spatial planning, ensuring the conservation of Brazil's unique rocky shore biodiversity and the ecosystem services they provide.

Acknowledgements We thank Dr. André Pardal for his valuable insights on the manuscript's content. This work was supported by the São Paulo Research Foundation (FAPESP: 2016/11947-7, A.S.M. 2022 Postdoctoral fellowship) and the National Council for Scientific and Technological Development (CNPq: 151903/2022-0, A.S.M. 2023 Postdoctoral fellowship).

Competing Interest The authors have no conflict of interests to declare that are relevant to the content of this chapter.

References

Abessa DMS, Zaroni LP, Sousa ECPM, Gasparro MR et al (2005) Physiological and cellular responses in two populations of the mussel *Perna perna* collected at different sites from the coast of São Paulo, Brazil. Braz Arch Biol Technol 48:217–225

Adelir-Alves J, Rocha GRA, Souza TF, Pinheiro PC et al (2016) Abandoned, lost or otherwise discarded fishing gears in rocky reefs of Southern Brazil. Braz J Oceanogr 64:427–434

Agostini VO, Ozorio CP (2016) Colonization record of *Isognomon bicolor* (Mollusca: Bivalvia) on pipeline monobuoys in the Brazilian south coast. Mar Biodivers Rec 9:1–4

Airoldi L, Turon X, Perkol-Finkel S, Rius M (2015) Corridors for aliens but not for natives: effects of marine urban sprawl at a regional scale. Divers Distrib 21:755–768. https://doi.org/10.1111/ddi.12301

Alencar JR (2010) Cultivo de camarão branco *Litopenaeus Vannamei* (Boone, 1931) com a macroalga *Ulva Lacuata Linneaus* (Chlorophyta) no tratamento de efluentes em sistema fechado de recirculação. Rev Biol Cienc Terra 10:117–137

Alves RRN, Dias TLP (2010) Usos de invertebrados na medicina popular no Brasil e suas implicações para conservação. Trop Conserv Sci 3:159–174. https://doi.org/10.1177/194008291000300204

Anderson A, Bonaldo R, Barneche D, Hackradt CW et al (2014) Recovery of grouper assemblages indicates effectiveness in a Marine Protected Area in Southern Brazil. Mar Ecol Prog Ser 514:207–215. https://doi.org/10.3354/meps11032

Andrade LR, Farina M, Amado GM (2004) Effects of copper on *Enteromorpha flexuosa* (Chlorophyta) in vitro. Ecotoxicol Environ Saf 58:117–125. https://doi.org/10.1016/s0147-6513(03)00106-4

Aragão JAN (2013) Pesca de lagostas no Brasil: Monitorar para ordenar. Boletin Técnico Científico CEPENE 19:103–106

Araújo ME, Mattos FMG, Melo FPL, Chaves LCT et al (2020) Diversity patterns of reef fish along the Brazilian tropical coast. Mar Environ Res 160:105038. https://doi.org/10.1016/j.marenvres.2020.105038

Assis JE, Alonso C, de Brito RJ, dos Santos AS et al (2012) Polychaetous annelids from the coast of Paraíba State, Brazil. Rev Nordestina Biol 21:3–45

Barneche DR, Anderson AB, Floeter SR, Silveira M et al (2009) Ten new records of reef fish on the coast of Santa Catarina State, Brazil. Mar Biodivers Rec 2:e143

Barradas JI, Chow F, Dias GM, Ghilardi-Lopes NP (2018) The influence of trampling disturbance on the fluorescence and pigment concentration of Sargassum beds (Fucales). J Exp Mar Biol Ecol 506:163–170

Barros MSF, dos Santos Calado TC (2020) Plastic ingestion lead to reduced body condition and modified diet patterns in the rocky shore crab Pachygrapsus transversus (Gibbes, 1850) (Brachyura: Grapsidae). Mar Pollut Bull 156:111249

Barros RC, Rocha RM, Pie MR (2009) Human-mediated global dispersion of *Styela plicata* (Tunicata, Ascidiacea). Aquat Invasions 4:45–57

Barros MSF, dos Santos Calado TC, Silva AS, dos Santos EV (2020) Ingestion of plastic debris affects feeding intensity in the rocky shore crab Pachygrapsus transversus Gibbes 1850 (Brachyura: Grapsidae). Int J Biodivers Conserv 12:113–117

Bastos N, Calazans SH, Altvater L, Neves EG et al (2022) Western Atlantic invasion of sun corals: incongruence between morphology and genetic delimitation among morphotypes in the genus Tubastraea. Bull Mar Sci 98:187–210

Baumann H, Doherty O (2013) Decadal changes in the world's coastal latitudinal temperature gradients. PLoS One 8:e67596. https://doi.org/10.1371/journal.pone.0067596

Baussant T, Ortiz-Zarragoitia M, Cajaraville MP, Bechmann RK et al (2011) Effects of chronic exposure to dispersed oil on selected reproductive processes in adult blue mussels (*Mytilus edulis*) and the consequences for the early life stages of their larvae. Mar Pollut Bull 62:1437–1445. https://doi.org/10.1016/j.marpolbul.2011.04.029

Begliomini FN, Maciel DC, de Almeida SM, Abessa DM et al (2017) Shell alterations in limpets as putative biomarkers for multi-impacted coastal areas. Environ Pollut 226:494–503. https://doi.org/10.1016/j.envpol.2017.04.045

Bishop MJ, Mayer-Pinto M, Airoldi L, Firth LB et al (2017) Effects of ocean sprawl on ecological connectivity: impacts and solutions. J Exp Mar Biol Ecol 492:7–30. https://doi.org/10.1016/j.jembe.2017.01.021

Borelli E (2007) Urbanização e qualidade ambiental: O processo de produção do espaço da costa brasileira. Rev Intern Interdisciplinar Interthesis 4:27

Bouzon ZL, Ferreira EC, dos Santos R, Scherner F et al (2012) Influences of cadmium on fine structure and metabolism of *Hypnea musciformis* (Rhodophyta, Gigartinales) cultivated in vitro. Protoplasma 249:637–650. https://doi.org/10.1007/s00709-011-0301-6

Breves-Ramos A, Junqueira AOR, Lavrado HP, Silva SHG et al (2010) Population structure of the invasive bivalve *Isognomon bicolor* on rocky shores of Rio de Janeiro State (Brazil). J Mar Biol Assoc UK 90:453–459. https://doi.org/10.1017/S0025315409990919

Bumbeer J, Rocha RM (2016) Invading the natural marine substrates: a case study with invertebrates in South Brazil. Zoologia 33:e20150211. https://doi.org/10.1590/S1984-4689zool-20150211

Buruaem LM, Hortellani MA, Sarkis JE, Costa-Lotufo LV et al (2012) Contamination of port zone sediments by metals from Large Marine Ecosystems of Brazil. Mar Pollut Bull 64:479–488. https://doi.org/10.1016/j.marpolbul.2012.01.017

Buschmann AH, Varela DA, Hernández-González MC, Huovinen P (2008) Opportunities and challenges for the development of an integrated seaweed-based aquaculture activity in Chile: determining the physiological capabilities of Macrocystis and Gracilaria as biofilters. J Appl Phycol 20:571–577. https://doi.org/10.1007/s10811-007-9297-x

Cairns SD (1994) Scleractinia of the temperate North Pacific. Smithsonian Institution Press, Washington, DC

Cairns S (2000) A revision of the shallow-water azooxanthellate Scleractinia of the Western Atlantic. Stud Nat Hist Caribb Reg 75:1–192

Capel KCC, Creed J, Kitahara M, Chen C et al (2019) Multiple introductions and secondary dispersion of Tubastraea spp. in the Southwestern Atlantic. Sci Rep 9:13978

Carlton JT (2009) Deep invasion ecology and the assembly of communities in hiistorical time. In: Rilov G, Crooks JA (eds) Biological invasions in marine ecosystems: ecological, management, and geographic perspectives. Springer, Berlin, Heidelberg, pp 13–56

Carlton JT, Ruiz GM (2005) Vector science and integrated vector management in bioinvasion ecology: conceptual frameworks. In: Invasive alien species: a new synthesis. Island Press, Washington, DC

Carranza A, Defeo O, Beck M, Castilla JC (2009) Linking fisheries management and conservation in bioengineering species: the case of South American mussels (Mytilidae). Rev Fish Biol Fish 19:349–366. https://doi.org/10.1007/s11160-009-9108-3

Castelar B, Reis RP, Bastos M (2009a) Contribuição ao protocolo de monitoramento ambiental da maricultura de *Kappaphycus alvarezii* (Doty) Doty ex P.C. Silva (Areschougiaceae – Rhodophyta) na baía de Sepetiba, RJ, Brasil. Acta Bot Bras 23:613–617

Castelar B, Reis RP, Moura AL, Kirk R (2009b) Invasive potential of *Kappaphycus alvarezii* off the south coast of Rio de Janeiro state, Brazil: a contribution to environmentally secure cultivation in the tropics. Bot Mar 52:283–289. https://doi.org/10.1515/BOT.2009.002

Castro CB, Pires DO (2001) Brazilian coral reefs: what we already know and what is still missing. Bull Mar Sci 69:357–371

Castro IB, Rocha-Barreira CD, Fernandez MA, Bigatti G (2012) Transplant bioassay induces different imposex responses in two species of the genus *Stramonita*. Mar Biol Res 8:397–404. https://doi.org/10.1080/17451000.2011.627923

Cesar-Ribeiro C, Palanch-Hans MF (2010) Chronic toxicity test with sea urchin *Echinometra lucunter* and *Lytechinus variegatus* (Echinodermata: Echinoidea), exposed to light-stick – flag paternoster used for longline surface fishing. Braz J Oceanogr 58:71–75

Cesar-Ribeiro C, Rosa HC, Rocha DO, dos Reis CGB et al (2017) Light-stick: a problem of marine pollution in Brazil. Mar Pollut Bull 117:118–123. https://doi.org/10.1016/j.marpolbul.2017.01.055

Chapman MG (2003) Paucity of mobile species on constructed seawalls: effects of urbanization on biodiversity. Mar Ecol Prog Ser 264:21–29. https://doi.org/10.3354/meps264021

Coelho SCC, Gherardi DFM, Gouveia MB, Kitahara MV (2022) Western boundary currents drive sun-coral (*Tubastraea* spp.) coastal invasion from oil platforms. Sci Rep 12:5286. https://doi.org/10.1038/s41598-022-09269-8

Coleman FC, Williams SL (2002) Overexploiting marine ecosystem engineers: potential consequences for biodiversity. Trends Ecol Evol 17:40–44. https://doi.org/10.1016/s0169-5347(01)02330-8

Corrêa TR, Calderon E, da Silva Lima L, de Freitas Delgado J et al (2023) Bioinvasion of benthic invertebrates in a eutrophicated tropical estuarine system in South America (Santos, Brazil). Environ Prot 3:38–55

Costa KG, Nalesso RC (2006) Effects of mussel farming on macrobenthic community structure in Southeastern Brazil. Aquaculture 258:655–663. https://doi.org/10.1016/j.aquaculture.2006.04.023

Costa GB, Simioni C, Pereira DT, Ramlov F et al (2017a) The brown seaweed Sargassum cymosum: changes in metabolism and cellular organization after long-term exposure to cadmium. Protoplasma 254:817–837. https://doi.org/10.1007/s00709-016-0992-9

Costa GB, Simioni C, Ramlov F, Maraschin M et al (2017b) Effects of manganese on the physiology and ultrastructure of Sargassum cymosum. Environ Exp Bot 133:24–34. https://doi.org/10.1016/j.envexpbot.2016.09.007

Costa-Paiva EM (2006) Estudo Taxonômico de *Branchiomma nigromaculatum* (Baird, 1865) (Annelida: Polychaeta: Sabellidae) na Costa Brasileira. Universidade Federal do Rio de Janeiro

Coutinho R, Yaginuma LE, Siviero F, Santos JCQP et al (2016) Studies on benthic communities of rocky shores on the Brazilian coast and climate change monitoring: status of knowledge and challenges. Braz J Oceanogr 64:27–36

Craveiro N, de Almeida Alves RV, da Silva JM, Vasconcelos E et al (2021) Immediate effects of the 2019 oil spill on the macrobenthic fauna associated with macroalgae on the tropical coast of Brazil. Mar Pollut Bull 165:112107. https://doi.org/10.1016/j.marpolbul.2021.112107

Creed JC (2006) Two invasive alien azooxanthellate corals, *Tubastraea coccinea* and *Tubastraea tagusensis*, dominate the native zooxanthellate Mussismilia hispida in Brazil. Coral Reefs 25:350–350

Cruz R, Santana JVM, Barreto CG, Borda CA et al (2020) Towards the rebuilding of spiny lobster stocks in Brazil: a review. Crustaceana 93:957–983. https://doi.org/10.1163/15685403-bja10073

Da Silva AG, de Paula AF, Fleury BG, Creed JC (2014) Eleven years of range expansion of two invasive corals (*Tubastraea coccinea* and *Tubastraeq tagusensis*) through the southwest Atlantic (Brazil). Estuar Coasts 141:9–16

Darrigran G, Agudo-Padrón I, Baez P, Belz C et al (2020) Non-native mollusks throughout South America: emergent patterns in an understudied continent. Biol Invasions 22:853–871. https://doi.org/10.1007/s10530-019-02178-4

de Oliveira Soares M, Salani S, Paiva SV, Braga MDA (2020) Shipwrecks help invasive coral to expand range in the Atlantic Ocean. Mar Pollut Bull 158:111394

de Paula AF, Creed JC (2004) Two species of the coral *Tubastraea* (Cnidaria, Scleractinia) in Brazil: a case of accidental introduction. Bull Mar Sci 74:175–183

de Santana CO, de Jesus TB, de Aguiar WM, Franca-Rocha W et al (2017) Assessment of health risk related to the ingestion of trace metals through fish consumption in Todos os Santos Bay. Environ Monit Assess 189:204. https://doi.org/10.1007/s10661-017-5913-7

Dias TLP, Leo Neto NA, Alves RRN (2011) Molluscs in the marine curio and souvenir trade in NE Brazil: species composition and implications for their conservation and management. Biodivers Conserv 20:2393–2405. https://doi.org/10.1007/s10531-011-9991-5

Dias PJ, Gilg MR, Lukehurst SS, Kennington WJ et al (2018) Genetic diversity of a hitchhiker and prized food source in the Anthropocene: the Asian green mussel *Perna viridis* (Mollusca, Mytilidae). Biol Invasions 20:1749–1770

Domaneschi O, Martins CM (2002) *Isognomon bicolor* (C.B. Adams) (Bivalvia, Isognomonidae): Primeiro registro para o Brasil, redescrição da espécie e considerações sobre a ocorrência e distribuição de Isognomon na costa brasileira. Rev Bras Zool 19:611

Duarte LFA, Pinheiro A, Santana W, Rodrigues E et al (2016) Avaliação das lagostas-sapateiras (Decapoda: Scyllaridae). In: Pinheiro M, Boss H (eds) Livro Vermelho dos Crustáceos do Brasil: Avaliação 2010–2014. Sociedade Brasileira de Carcinologia, Porto Alegre, pp 377–389

Durante LM, Cruz ICS, Lotufo TMC (2018) The effect of climate change on the distribution of a tropical zoanthid (Palythoa caribaeorum) and its ecological implications. PeerJ 6:e4777. https://doi.org/10.7717/peerj.4777

Eggertsen L, Luza AL, Cordeiro CAMM, Dambros C et al (2024) Complexities of reef fisheries in Brazil: a retrospective and functional approach. Rev Fish Biol Fish 34:511–538. https://doi.org/10.1007/s11160-023-09826-y

Elliott M (2003) Biological pollutants and biological pollution—an increasing cause for concern. Mar Pollut Bull 46:275–280. https://doi.org/10.1016/S0025-326X(02)00423-X

Erlandson JM, Rick TC (2009) Archaeology meets marine ecology: the antiquity of maritime cultures and human impacts on marine fisheries and ecosystems. Annu Rev Mar Sci 2:231–251. https://doi.org/10.1146/annurev.marine.010908.163749

Falcão C, Széchy MTM (2005) Changes in shallow phytobenthic assemblages in southeastern Brazil, following the replacement of *Sargassum vulgare* (Phaeophyta) by *Caulerpa scalpelliformis* (Chlorophyta). Bot Mar 48:208–217. https://doi.org/10.1515/BOT.2005.033

FAO (2007) Estado actual del cultivo y manejo de moluscos bivalvos y su proyección futura: Factores que afectan su sustentabilidad en América Latina. Taller Técnico Regional de la FAO, 20–24 de agosto de 2007, Puerto Montt, Chile. FAO Actas de Pesca y Acuicultura, 12, Rome, Italy

Felix MRD, Osorio LKP, Ouriques LC, Farias-Soares FL et al (2014) The effect of cadmium under different salinity conditions on the cellular architecture and metabolism in the red alga *Pterocladiella capillacea* (Rhodophyta, Gelidiales). Microsc Microanal 20:1411–1424. https://doi.org/10.1017/s1431927614012768

Ferreira C, Gonçalves J (1999) The unique Abrolhos reef formation (Brazil): need for specific management strategies. Coral Reefs 18:352–352. https://doi.org/10.1007/s003380050211

Ferreira M, Rosso S (2009) Effects of human trampling on a rocky shore fauna on the Sao Paulo coast, southeastern Brazil. Braz J Biol 69:993–999

Fistarol GO, Coutinho FH, Moreira APB, Venas T et al (2015) Environmental and sanitary conditions of Guanabara Bay, Rio de Janeiro. Front Microbiol 6:article 1232. https://doi.org/10.3389/fmicb.2015.01232

Floeter SR, Halpern BS, Ferreira CEL (2006) Effects of fishing and protection on Brazilian reef fishes. Biol Conserv 128:391–402. https://doi.org/10.1016/j.biocon.2005.10.005

Francioni E, Wagener AD, Scofield AD, Depledge MH et al (2007) Polycyclic aromatic hydrocarbon in inter-tidal mussel *Perna perna*: space-time observations, source investigation and genotoxicity. Sci Total Environ 372:515–531. https://doi.org/10.1016/j.scitotenv.2006.08.046

Frédou T, Ferreira BP, Letourneur Y (2006) A univariate and multivariate study of reef fisheries off northeastern Brazil. ICES J Mar Sci 63:883–896. https://doi.org/10.1016/j.icesjms.2005.11.019

Garcia SM, Du Clos KT, Hawkins OH, Gemmell BJ (2020) Sublethal effects of crude oil and chemical dispersants on multiple life history stages of the Eastern oyster, *Crassostrea virginica*. J Mar Sci Eng 8:0808

Gasparini JL, Floeter SR, Ferreira CEL, Sazima I (2005) Marine ornamental trade in Brazil. Biodivers Conserv 14:2883–2899. https://doi.org/10.1007/s10531-004-0222-1

Giglio VJ, Bender MG, Zapelini C, Ferreira CEL (2017) The end of the line? Rapid depletion of a large-sized grouper through spearfishing in a subtropical marginal reef. Perspect Ecol Conserv 15:115–118. https://doi.org/10.1016/j.pecon.2017.03.006

Giglio VJ, Ternes ML, Luiz OJ, Zapelini C et al (2018) Human consumption and popular knowledge on the conservation status of groupers and sharks caught by small-scale fisheries on Abrolhos Bank, SW Atlantic. Mar Policy 89:142–146

Gittman RK, Scyphers SB, Smith CS, Neylan IP et al (2016) Ecological consequences of shoreline hardening: a meta-analysis. Bioscience 66:763–773. https://doi.org/10.1093/biosci/biw091

Gorman D, Horta P, Flores AAV, Turra A et al (2020) Decadal losses of canopy-forming algae along the warm temperate coastline of Brazil. Glob Chang Biol 26:1446–1457. https://doi.org/10.1111/gcb.14956

Gouveia C, Kreusch M, Schmidt EC, Felix MRD et al (2013) The effects of lead and copper on the cellular architecture and metabolism of the red alga *Gracilaria domingensis*. Microsc Microanal 19:513–524. https://doi.org/10.1017/s1431927613000317

Grande H, Reis M, Jacobucci GB (2012) Small-scale experimental contamination with diesel oil does not affect the recolonization of *Sargassum* (Fucales) fronds by vagile macrofauna. Zoologia 29:135–143. https://doi.org/10.1590/s1984-46702012000200006

Granthom-Costa LV, Resende de Messano LV, Padula V, Silva Oliveira FA et al (2023) First record of a *Didemnum* carpet ascidian from the southwestern Atlantic Ocean. Bioinvasions Rec 12:753

Grosholz E (2002) Ecological and evolutionary consequences of coastal invasions. Trends Ecol Evol 17:22–27

Guabiroba HC, Santos MEA, Pinheiro HT, Simon T et al (2020) Trends in recreational fisheries and reef fish community structure indicate decline in target species population in an isolated tropical oceanic island. Ocean Coast Manag 191:105194

Gurjão LM, Lotufo TMC (2018) Native species exploited by marine aquarium trade in Brazil. Biota Neotrop 18:e20170387

Gurjão LM, Barros GM, Lopes DP, Machado DA et al (2017) Illegal trade of aquarium species through the Brazilian postal service in Ceará State. Mar Freshw Res 69:178–185

Gusmao JB, Albergaria-Barbosa ACR, Kikuchi RKP, Combi T (2021) The barnacle *Chthamalus bisinuatus* is the only sessile invertebrate colonizing oil patches on beachrocks one year after a massive oil spill on the Northeastern Brazilian coast. Mar Pollut Bull 173:112952. https://doi.org/10.1016/j.marpolbul.2021.112952

Haimovici M, Andriguetto Filho JM, Sunye PS (2014a) A pesca marinha e estuarina no Brasil. Universidade Federal do Rio Grande

Haimovici M, Leite T, Marinho R, Batista B et al (2014b) As pescarias de polvos do Nordeste do Brasil. In: Haimovici M, Filho JMA, Sunye PS (eds) A pesca marinha e estuarina no Brasil: Estudos de caso multidisciplinares. Editora da FURG, Rio Grande, pp 147–160

Hardison A, Anderson IC, Canuel EA, Tobias C et al (2011) Carbon and nitrogen dynamics in shallow photic systems: interactions between macroalgae, microalgae, and bacteria. Limnol Oceanogr 56:1489–1503

Hatje V, Pedreira RMA, de Rezende CE, Schettini CAF et al (2017) The environmental impacts of one of the largest tailing dam failures worldwide. Sci Rep 7:10706. https://doi.org/10.1038/s41598-017-11143-x

Hayashi L, Reis RP (2012) Cultivation of the red algae *Kappaphycus alvarezii* in Brazil and its pharmacological potential. Rev Bras 22:748–752

Henriques MB, de Almeida Marques HL, Barrella W, Pereira OM (2001) Estimativa do tempo de recuperação de um banco natural do mexilhão *Perna perna* (Linnaeus, 1758) na Baía de Santos, Estado de São Paulo. Holos Environ 1:85–100

Henriques M, Marques HLA, Pereira O, Bastos G (2004) Aspectos da estrutura populacional do mexilhão *Perna perna*, relacionados à extração em bancos naturais da Baía de Santos, estado de São Paulo, Brasil. Bol Inst Pesca 30:117–126

IBAMA (2020a) Instituto Brasileiro do Meio Ambiente e dos Recursos Naturais Renováveis. Localidades afetadas. Available in https://www.ibama.gov.br/phocadownload/emergenciasambientais/2020/manchasdeoleo/2020-03-19_LOCALIDADES_AFETADAS.pdf

IBAMA (2020b) Plano Nacional de Prevenção, Controle e Monitoramento do Coral-sol (Tubastraea spp.) no Brasil. Instituto Brasileiro do Meio Ambiente e dos Recursos Naturais Renováveis (IBAMA). Diretoria de Uso Sustentável da Biodiversidade e Florestas, Brasília, DF, p 113

IBGE (2023) Instituto Brasileiro de Geografia e Estatística. Censo Demográfico 2022. https://censo2022.ibge.gov.br/panorama/. Last acess: August 2023

ICMBio/MMA (2018) Livro Vermelho da Fauna Brasileira Ameaçada de Extinção: Volume I. ICMBio/MMA, Brasília

Instituto da Pesca (2017) Sobre a pesca e o homem – a pesca no Brasil. https://institutodepesca.org/ip-na-midia/249-sobre-a-pesca-e-o-homem-a-pesca-no-brasil. Acessado em 05/02/2021

Jackson JBC, Kirby MX, Berger WH, Bjorndal KA et al (2001) Historical overfishing and the recent collapse of coastal ecosystems. Science 293:629–637. https://doi.org/10.1126/science.1059199

Klautau M, Monteiro L, Borojevic R (2004) First occurrence of the genus *Paraleucilla* (Calcarea, Porifera) in the Atlantic Ocean: *P. magna* sp. nov. Zootaxa 710:1–8

Klautau M, Cóndor-Luján B, Azevedo F, Leocorny P et al (2020) *Heteropia glomerosa* (Bowerbank, 1873) (Porifera, Calcarea, Calcaronea), a new alien species in the Atlantic. Syst Biodivers 18:362–376

Kuhnen VV, de Góis Costa L, de Lima Raiol K, de Sousa OM et al (2019) Mariculture impacts on the benthonic icthyofauna of Itaguá bay, Ubatuba, southeast Brazil. Bol Inst Pesca 45:e500. https://doi.org/10.20950/1678-2305.2019.45.4.500

Lage H, Jablonski S (2008) Mussel *Perna perna* extraction and commercialization in Guanabara Bay, Brazil. Atlantica 30:161–169

Lages BG, Fleury BG, Menegola C, Creed JC (2011) Change in tropical rocky shore communities due to an alien coral invasion. Mar Ecol Prog Ser 438:85–96

Lamparelli ML, Costa MP, Prósperi VA, Bevilácqua JE et al (2001) Sistema estuarino de Santos e São Vicente. Technical report CETESB, São Paulo, p 178

Leal I, Flores AAV, Archambault P, Collin R et al (2020) Response of tropical and subtropical chthamalid barnacles to increasing substrate temperatures. J Exp Mar Biol Ecol 524:151281. https://doi.org/10.1016/j.jembe.2019.151281

Leite MBNL, de Araujo MMS, Nascimento IA, da Cruz ACS et al (2011) Toxicity of water-soluble fractions of biodiesel fuels derived from castor oil, palm oil, and waste cooking oil. Environ Toxicol Chem 30:893–897. https://doi.org/10.1002/etc.444

Leite LG, Ciotti ÁM, Christofoletti RA (2012) Abundance of biofilm on intertidal rocky shores: can trampling by humans be a negative influence? Mar Environ Res 79:111–115

Lima LLC, Passos FD (2021) Marine bioinvasions: bivalve molluscs introduced in northeast Brazil. Diversitas J 6:507–526

Link J, Segal B, Casarini LM (2019) Abandoned, lost or otherwise discarded fishing gear in Brazil: a review. Perspect Ecol Conserv 17:1–8

Lira ALO, Craveiro N, da Silva FF, Rosa Filho JS (2021) Effects of contact with crude oil and its ingestion by the symbiotic polychaete Branchiosyllis living in sponges (Cinachyrella sp.) following the 2019 oil spill on the tropical coast of Brazil. Sci Total Environ 801:149655. https://doi.org/10.1016/j.scitotenv.2021.149655

Loebmann D, Mai ACG, Lee JT (2010) The invasion of five alien species in the Delta do Parnaíba Environmental Protection Area, Northeastern Brazil. Rev Biol Trop 58:909–923

Lopes R, Villac M (2009) Informe sobre as espécies exóticas invasoras marinhas no Brasil. Ministério do Meio Ambiente – Secretaria de biodiversidade e Florestas, Brasília, pp 19–28

Lopez MS, Coutinho R, Ferreira CEL, Rilov G (2010) Predator prey interactions in a bioinvasion scenario: differential predation by native predators on two exotic rocky intertidal bivalves. Mar Ecol Prog Ser 403:101–112

Lopez MS, Lavrado HP, Coutinho R (2014) Structure of intertidal sessile communities before and after the invasion of *Isognomon bicolor* (C.B. Adams, 1845) (Bivalvia, Isognomonidae) in southeastern Brazil. Aquat Invasions 9:457–465

Luiz-Júnior O, Floeter S, Gasparini J, Ferreira C et al (2004) The occurrence of *Acanthurus monroviae* (Perciformes: Acanthuridae) in the south-western Atlantic, with comments on other eastern Atlantic reef fishes occurring in Brazil. J Fish Biol 65:1173–1179

Madeira C, Mendonça V, Flores AA, Diniz MS et al (2018a) High thermal tolerance does not protect from chronic warming – a multiple end-point approach using a tropical gastropod, *Stramonita haemastoma*. Ecol Indic 91:626–635

Madeira C, Mendonça V, Leal MC, Flores AA et al (2018b) Environmental health assessment of warming coastal ecosystems in the tropics – application of integrative physiological indices. Sci Total Environ 643:28–39

Madeira C, Mendonca V, Leal MC, Diniz MS et al (2019) Present and future invasion perspectives of an alien shrimp in South Atlantic coastal waters: an experimental assessment of functional biomarkers and thermal tolerance. Biol Invasions 21:1567–1584

Mantelatto MC, Creed JC, Mourão G, Migotto AE et al (2011) Range expansion of the invasive corals *Tubastraea coccinea* and *Tubastraea tagusensis* in the Southwest Atlantic. Coral Reefs 30:397–397

Manzoni G, Lacava L (1998) Crescimento dos gastrópodes *Thais* (*Stramonita*) *haemastoma* e *Cymatium parthenopeum parthenopeum* em cultivo experimental na enseada da armação do Itapocoroy (26O 47 S-48O 36 W) (Penha-SC). Braz J Aquat Sci Technol 2:167–173

Marinho-Soriano E (2017) Historical context of commercial exploitation of seaweeds in Brazil. J Appl Phycol 29:665–671

Marins FO, Novaes RLM, Rocha RM, Junqueira AOR (2010) Non indigenous ascidians in port and natural environments in a tropical Brazilian bay. Zoologia 27:213–221. https://doi.org/10.1590/s1984-46702010000200009

Martinez AS (2012) Spatial distribution of the invasive bivalve *Isognomon bicolor* on rocky shores of Arvoredo Island (Santa Catarina, Brazil). J Mar Biol Assoc UK 92:495–503. https://doi.org/10.1017/S0025315411000865

Martinez AS, Mayer-Pinto M, Christofoletti RA (2019) Functional responses of filter feeders increase with elevated metal contamination: are these good or bad signs of environmental health? Mar Pollut Bull 149:110571. https://doi.org/10.1016/j.marpolbul.2019.110571

Martinez AS, Underwood T, Christofoletti RA, Pardal A et al (2022) Reviewing the effects of contamination on the biota of Brazilian coastal ecosystems: scientific challenges for a developing country in a changing world. Sci Total Environ 803:150097. https://doi.org/10.1016/j.scitotenv.2021.150097

Martinez AS, Rocha RM, Christofoletti RA, Vieira LM et al (under review) Hardening of the Brazilian coast: ecological research and guidance for sustainability on the longest coastline in the South Atlantic

Martins CDL, Arantes N, Faveri C, Batista MB et al (2012) The impact of coastal urbanization on the structure of phytobenthic communities in southern Brazil. Mar Pollut Bull 64:772–778. https://doi.org/10.1016/j.marpolbul.2012.01.031

Mattos SMG, Wojciechowski MJ, Macnaughton AE, da Silva GHG et al (2017) Implementing the small-scale fisheries guidelines: lessons from Brazilian clam fisheries. The small-scale fisheries guidelines: Global implementation. Illuminating Hidden Harvests (IHH) Project, organized and coordinated by the Food and Agriculture Organization of the United Nations (FAO), the WorldFish, and the University of Duke, Instituto Maramar para a Gestão Responsável dos Ambientes Costeiros e Marinhos (Maramar Institute for Coastal Management). Brazil, pp 473–494

Mattos S, Wojciechowski M, Gandini F (2020) Iluminando as capturas ocultas da pesca artesanal costeira no Brasil: um estudo de caso. Relatório Executivo. Illuminating Hidden Harvests (IHH) Project, organized and coordinated by the Food and Agriculture Organization of the United Nations (FAO), the WorldFish, and the University of Duke, Instituto Maramar para a Gestão Responsável dos Ambientes Costeiros e Marinhos (Maramar Institute for Coastal Management). Brazil, p 71

Mazzuco ACA, Christofoletti RA, Pineda J, Starczak VR et al (2015) Temporal variation in intertidal community recruitment and its relationships to physical forcings, chlorophyll-a concentration and sea surface temperature. Mar Biol 162:1705–1725. https://doi.org/10.1007/s00227-015-2689-6

Mazzuco ACA, Stelzer PS, Donadia G, Bernardino JV et al (2019) Lower diversity of recruits in coastal reef assemblages are associated with higher sea temperatures in the tropical South Atlantic. Mar Environ Res 148:87–98. https://doi.org/10.1016/j.marenvres.2019.05.008

Medeiros RJ, dos Santos LMG, Freire AS, Santelli RE et al (2012) Determination of inorganic trace elements in edible marine fish from Rio de Janeiro State, Brazil. Food Control 23:535–541. https://doi.org/10.1016/j.foodcont.2011.08.027

Melo CMR, Silva FC, Gomes CHAM, Solé-Cava AM et al (2010) *Crassostrea gigas* in natural oyster banks in southern Brazil. Biol Invasions 12:441–449. https://doi.org/10.1007/s10530-009-9475-7

Menezes N, McFadden C, Miranda R, Nunes J et al (2021) New non-native ornamental octocorals threatening a South-west Atlantic reef. J Mar Biol Assoc UK 101:911–917

Messano L, Gonçalves JEA, Messano HF, Campos SHC et al (2019) First report of the Asian green mussel *Perna viridis* (Linnaeus, 1758) in Rio de Janeiro, Brazil: a new record for the southern Atlantic Ocean. Bioinvasions Rec 8:653–660

Miranda RJ, Cruz IC, Barros F (2016) Effects of the alien coral *Tubastraea tagusensis* on native coral assemblages in a southwestern Atlantic coral reef. Mar Biol 163:1–12

Moles P, Bunge J (2002) Shrimp farming in Brazil: an industry overview. Report prepared under the World Bank, NACA, WWF and FAO Consortium Program on Shrimp Farming and the Environment. Work in Progress for Public Discussion. Published by the Consortium

Monniot C, Monniot F (1997) Records of ascidians from Bahrain, Arabian Gulf with three new species. J Nat Hist 31:1623–1643

Monteiro LS (2011) Absorção de nutrientes pela macroalga Gracilaria Birdiae (Plastino & Oliveira, 2002) sob diferentes condições físicoquímicas. Departamento de Engenharia de Pesca, Universidade Federal do Ceará, p 84

Monteiro-Neto C, De Andrade Cunha FE, Carvalho Nottingham M, Araújo ME et al (2003) Analysis of the marine ornamental fish trade at Ceará State, northeast Brazil. Biodivers Conserv 12:1287–1295

Moraes CAR (2007) Contribuição para a gestão da zona costeira do Brasil: Elementos para uma geografia do litoral brasileiro. Annablume, São Paulo

MPA/Brasil (2010) Boletim Estatístico da Pesca e Aquicultura 2008 e 2009. Brasília, Ministério da Pesca e Aquicultura, p 101

Neto JBG, Goyanna FAA, Feitosa CV, Soares MO (2021) A sleeping giant: the historically neglected Brazilian fishing sector. Ocean Coast Manag 209:105699. https://doi.org/10.1016/j.ocecoaman.2021.105699

Netto SA, Valgas I (2010) The response of nematode assemblages to intensive mussel farming in coastal sediments (Southern Brazil). Environ Monit Assess 162:81–93. https://doi.org/10.1007/s10661-009-0777-0

Nobre CR, Santana MFM, Maluf A, Cortez FS et al (2015) Assessment of microplastic toxicity to embryonic development of the sea urchin *Lytechinus variegatus* (Echinodermata: Echinoidea). Mar Pollut Bull 92:99–104. https://doi.org/10.1016/j.marpolbul.2014.12.050

Nogueira JMM, Rossi MCS, López E (2006) Intertidal species of *Branchiomma Kolliker* and *Pseudobranchiomma Jones* (Polychaeta: Sabellidae: Sabellinae) occurring on rocky shores along the state of Sao Paulo, southeastern Brazil. Zool Stud 45:586

Nottingham MC, Cunha FEA, Monteiro Neto C (2000) Captura de peixes ornamentais marinhos no estado do Ceará. Arq Cienc Mar 33:113–118

Nottingham M, Barreto L, Araújo M, Monteiro-Neto C et al (2005) A explotação de peixes ornamentais marinhos no estado do Ceará, Brasil: Captura, manutenção nas empresas e exportação. Boletin Técnico Científico CEPENE 13:53–73

Nunes JACC, Medeiros DV, Reis-Filho JA, Sampaio CLS et al (2012) Reef fishes captured by recreational spearfishing on reefs of Bahia State, northeast Brazil. Biota Neotrop 12:179

Ohayashi NS, Rodrigues ID, Marchetti OC, Dias GM (2022) Seeding artificial habitats with native benthic species can prevent the occurrence of exotic organisms. Mar Environ Res 182:105771. https://doi.org/10.1016/j.marenvres.2022.105771

Oricchio FT, Marques AC, Hajdu E, Pitombo FB et al (2019) Exotic species dominate marinas between the two most populated regions in the southwestern Atlantic Ocean. Mar Pollut Bull 146:884–892. https://doi.org/10.1016/j.marpolbul.2019.07.013

Oshiro LMY (1999) Aspectos reprodutivos do caranguejo guaia, *Menippe nodifrons* Stimpson (Crustacea, Decapoda, Xanthidae) da Baía de Sepetiba, Rio de Janeiro, Brasil. Rev Bras Zool 16:827–834

Pardal A, Jenkins SR, Christofoletti RA (2022) Local and large-scale spatial variation in a marine predator-prey interaction in the southwestern Atlantic. Oecologia 199:685–698

Pardal A, Martinez AS, Ciotti ÁM, Christofoletti RA et al (2023) Macroecology of rocky intertidal benthic communities along the southwestern Atlantic: patterns of spatial variation and associations with natural and anthropogenic variables. Mar Environ Res 190:106099. https://doi.org/10.1016/j.marenvres.2023.106099

Pardal A, Christofoletti RA, Martinez AS (2024) Urbanisation on the coastline of the most populous and developed state of Brazil: the extent of coastal hardening and occupations in low-elevation zones. Anthr Coasts 7:15

Pardal-Souza AL, Dias GM, Jenkins SR, Ciotti ÁM et al (2017) Shading impacts by coastal infrastructure on biological communities from subtropical rocky shores. J Appl Ecol 54:826. https://doi.org/10.1111/1365-2664.12811

Pellizzari F, Reis RP (2011) Seaweed cultivation on the Southern and Southeastern Brazilian Coast. Rev Bras 21:305–312

Pereira CDS, Maranho LA, Cortez FS, Pusceddu FH et al (2016) Occurrence of pharmaceuticals and cocaine in a Brazilian coastal zone. Sci Total Environ 548-549:148–154. https://doi.org/10.1016/j.scitotenv.2016.01.051

Pereira TM, Mattar LP, Pereira ER, Mercon J et al (2017) Effects of tin on enzyme activity in *Holothuria grisea* (Echinodermata: Holothuroidea). Bull Environ Contam Toxicol 98:607–611. https://doi.org/10.1007/s00128-016-1992-3

Pereira PHC, Ternes MLF, Nunes JAC, Giglio VJ (2021) Overexploitation and behavioral changes of the largest South Atlantic parrotfish (*Scarus trispinosus*): evidence from fishers' knowledge. Biol Conserv 254:108940

Pilatti FK, Ramlov F, Schmidt EC, Kreusch M et al (2016) In vitro exposure of *Ulva lactuca* Linnaeus (Chlorophyta) to gasoline – biochemical and morphological alterations. Chemosphere 156:428–437. https://doi.org/10.1016/j.chemosphere.2016.04.126

Pineda MC, Lopez-Legentil S, Turon X (2011) The whereabouts of an ancient wanderer: global phylogeography of the solitary ascidian *Styela plicata*. PLoS One 6:e25495

Pinheiro HT, Joyeux J-C, Martins AS (2010) Reef fisheries and underwater surveys indicate overfishing of a Brazilian coastal island. Nat Conserv 8:151–159

Pinheiro A, Alencar CER, Freire FA (2016) Avaliação do caranguejo-Guajá *Carpilius corallinus* (Herbst, 1783) (DECAPODA: CARPILIIDAE). In: Pinheiro M, Boos H (eds) Livro Vermelho dos Crutáceos do Brasil: Avaliação 2010–2014. Sociedade Brasileira de Carcinologia, Porto Alegre, pp 126–134

Pinheiro FCF, Pinheiro HT, Costa TJF, Teixeira JB et al (2018) Harvest of endangered marine invertebrates in a priority area for conservation in Brazil. Nat Conserv Res 3:78–81

Queiroz RNM, Dias TLP, Batista R, da Silva PM (2022) Reproduction and population dynamics of the invasive bivalves Mytilopsis sallei and Isognomon bicolor on the Northeast coast of Brazil. Zoology 153:126028

Queiroz R, Barros G, Oliveira L, Avelino C et al (2023) Influence of the invasive bivalve *Isognomon bicolor* on the native community found in northeast Brazilian coast. Can J Zool 101:551–559. https://doi.org/10.1139/cjz-2022-0138

Queiroz TC, Yokoyama LQ, Dias GM (2024) Does the incorporation of shell waste from aquaculture in the construction of marine facilities affect the structure of the marine sessile community? Mar Environ Res 198:106484. https://doi.org/10.1016/j.marenvres.2024.106484

Rich WA, Schubert N, Schläpfer N, Carvalho VF et al (2018) Physiological and biochemical responses of a coralline alga and a sea urchin to climate change: implications for herbivory. Mar Environ Res 142:100–107

Rocha RM, Kremer LP (2005) Introduced ascidians in Paranaguá Bay, Paraná, southern Brazil. Zoologia 22:1170–1184

Rocha RM, Kremer LP, Baptista MS, Metri R (2009) Bivalve cultures provide habitat for exotic tunicates in southern Brazil. Aquat Invasions 4:195–205

Rola RC, Monteiro MD, Reis SRD, Sandrini JZ (2012) Molecular and biochemical biomarkers responses in the mussel *Mytilus edulis* collected from Southern Brazil coast. Mar Pollut Bull 64:766–771. https://doi.org/10.1016/j.marpolbul.2012.01.004

Rolim FA, Langlois T, Rodrigues PF, Bond T et al (2019) Network of small no-take marine reserves reveals greater abundance and body size of fisheries target species. PLoS One 14:e0204970

Rosa Filho J, Pereira L, Aviz D, Braga C et al (2018) Benthic estuarine assemblages of the Brazilian north coast (Amazonia ecoregion): a benthic perspective. In: Brazilian estuaries. Springer, Cham, pp 39–74

Ruiz GM, Carlton JT (2003) Invasive species vectors and management strategies. Island Press, London

Rupp GS, Parsons GJ (2006) Scallop aquaculture and fisheries in Brazil. In: Developments in aquaculture and fisheries science. Elsevier, Amsterdam, pp 1225–1250

Sanderson EW, Jaiteh M, Levy MA, Redford KH et al (2002) The human footprint and the last of the wild: the human footprint is a global map of human influence on the land surface, which suggests that human beings are stewards of nature, whether we like it or not. Bioscience 52:891–904. https://doi.org/10.1641/0006-3568

Santana W, Ivo CT, Neto JD, Duarte LFA et al (2010) Avaliação das lagostas-de-espinho (Decapoda: Palinuridae). In: Pinheiro M, Boss H (eds) Livro vermelho dos crustáceos do Brasil: Avaliação. Sociedade Brasileira de Carcinologia, Porto Alegre, pp 268–283

Santana JL, Rios AS, Calado TCS, Zanardi-Lamardo E et al (2022) Reef crab population changes after oil spill disaster reach Brazilian tropical environments. Mar Pollut Bull 183:114047. https://doi.org/10.1016/j.marpolbul.2022.114047

Santos RW, Schmidt EC, Vieira IC, Costa GB et al (2015) The effect of different concentrations of copper and lead on the morphology and physiology of *Hypnea musciformis* cultivated in vitro: a comparative analysis. Protoplasma 252:1203–1215. https://doi.org/10.1007/s00709-014-0751-8

Santos HS, Bertollo JC, Creed JC (2023) Range extension of the Asian green mussel *Perna viridis* (Linnaeus, 1758) into a Marine Extractive Reserve in Brazil. Bioinvasions Rec 12:208–222

Scherner F, Barufi JB, Horta PA (2012) Photosynthetic response of two seaweed species along an urban pollution gradient: evidence of selection of pollution-tolerant species. Mar Pollut Bull 64:2380–2390. https://doi.org/10.1016/j.marpolbul.2012.08.012

Schmidt EC, Felix MRD, Polo LK, Kreusch MG et al (2015) Influence of cadmium and salinity in the red alga *Pterocladiella capillacea*: cell morphology, photosynthetic performance and antioxidant systems. Braz J Bot 38:737–749. https://doi.org/10.1007/s40415-015-0183-5

Silva e Cavalcante L, Gomes ATA (2022) Population structure of the invasive bivalve *Isognomon bicolor* (CB Adams, 1845) on rocky beaches in northeastern Brazil. Stud Environ Anim Sci 3:245–257

Silva JN, Ghilardi-Lopes NP (2012) Indicators of the impacts of tourism on hard-bottom benthic communities of Ilha do Cardoso State Park (Cananéia) and Sonho Beach (Itanhaém), two southern coastal areas of São Paulo State (Brazil). Ocean Coast Manag 58:1–8

Silva EJ, Martins IX (2017) A pesca de moluscos em ambientes intermareais no oeste do Estado do Rio Grande do Norte, Brasil. Arq Cienc Mar 50:110–118

Silva BNT, Amancio CE, Oliveira-Filho EC (2010) Exotic marine macroalgae on the Brazilian coast: a revision. Oecol Aust 14:403–414

Silva R, Vinagre C, Kitahara MV, Acorsi IV et al (2019) Sun coral invasion of shallow rocky reefs: effects on mobile invertebrate assemblages in Southeastern Brazil. Biol Invasions 21:1339–1350. https://doi.org/10.1007/s10530-018-1903-0

Silvano RA, Nora V, Andreoli TB, Lopes PF et al (2017) The "ghost of past fishing": small-scale fisheries and conservation of threatened groupers in subtropical islands. Mar Policy 75:125–132

Simioni C, Hayashi L, Oliveira MC (2019) Seaweed resources of Brazil: what has changed in 20 years? Bot Mar 62:433–441

Soares MO, Davis M, de Macêdo Carneiro PB (2016) Northward range expansion of the invasive coral (*Tubastraea tagusensis*) in the southwestern Atlantic. Mar Biodivers 48:1651–1654. https://doi.org/10.1007/s12526-016-0623-x

Szalaj D, De Orte MR, Goulding TA, Medeiros ID et al (2017) The effects of ocean acidification and a carbon dioxide capture and storage leak on the early life stages of the marine mussel Perna perna (Linneaus, 1758) and metal bioavailability. Environ Sci Pollut Res Int 24:765–781. https://doi.org/10.1007/s11356-016-7863-y

Tanasovici RM, Kitahara MV, Dias GM (2020) Invasive coral *Tubastraea* spp. population growth in artificial habitats and its consequences to the diversity of benthic organisms. Mar Biol 167:119. https://doi.org/10.1007/s00227-020-03734-6

Teixeira L, Creed J (2020) A decade on: an updated assessment of the status of marine non-indigenous species in Brazil. Aquat Invasions 15:30–43. https://doi.org/10.3391/ai.2020.15.1.03

Valenti WC, Barros HP, Moraes-Valenti P, Bueno GW et al (2021) Aquaculture in Brazil: past, present and future. Aquac Res 19:100611. https://doi.org/10.1016/j.aqrep.2021.100611

Vasconcelos MA, Schubart CLQ, Széchy MTM (2011) Temporal variation in vegetative development of *Caulerpa scalpelliformis* (Chlorophyta) from Baleia beach, Ilha Grande bay (Rio de janeiro, Brazil). Braz J Oceanogr 59:145–152

Vasconcelos ERTPP, Vasconcelos JB, Reis TNV, Cocentino ALM et al (2019) Macroalgal responses to coastal urbanization: relative abundance of indicator species. J Appl Phycol 31:893–903

Ventura CR, Borges M, Campos LS, Costa-Lotufo LV et al (2013) Echinoderm from Brazil: historical research and the current state of biodiversity knowledge. In: Echinoderm research and diversity in Latin America. Springer, Berling, Heidelberger, pp 301–344

Vergés A, Steinberg PD, Hay ME, Poore AGB et al (2014) The tropicalization of temperate marine ecosystems: climate-mediated changes in herbivory and community phase shifts. Proc R Soc B 281:20140846. https://doi.org/10.1098/rspb.2014.0846

Vignier J, Volety AK, Rolton A, Le Goïc N et al (2017) Sensitivity of eastern oyster (*Crassostrea virginica*) spermatozoa and oocytes to dispersed oil: cellular responses and impacts on fertilization and embryogenesis. Environ Pollut 225:270–282. https://doi.org/10.1016/j.envpol.2016.11.052

Vinagre C, Leal I, Mendonca V, Madeira D et al (2016) Vulnerability to climate warming and acclimation capacity of tropical and temperate coastal organisms. Ecol Indic 62:317–327

Vinagre C, Mendonca V, Cereja R, Abreu-Afonso F et al (2018) Ecological traps in shallow coastal waters—potential effect of heat-waves in tropical and temperate organisms. PLoS One 13:e0192700

Vinagre C, Dias M, Cereja R, Abreu-Afonso F et al (2019) Upper thermal limits and warming safety margins of coastal marine species – indicator baseline for future reference. Ecol Indic 102:644–649

Vitousek PM, Mooney HA, Lubchenco J, Melillo JM (1997) Human domination of earth's ecosystems. Science 277:494–499

Williams SL, Davidson IC, Pasari JR, Ashton GV et al (2013) Managing multiple vectors for marine invasions in an increasingly connected world. Bioscience 63:952–966

Wood EM (2001) Collection of coral reef fish for aquaria: global trade, conservation issues and management strategies. Marine Conservation Society, Ross-on-Wye, p 80

Wu L, Cai W, Zhang L, Nakamura H et al (2012) Enhanced warming over the global subtropical western boundary currents. Nat Clim Chang 2:161–166. https://doi.org/10.1038/nclimate1353

Xavier EA, Almeida ACS, Vieira LM (2021) The role of artificial habitats on fouling bryozoan fauna in the southwestern Atlantic. Mar Pollut Bull 167:112310. https://doi.org/10.1016/j.marpolbul.2021.112310

Zalmon IR, Krohling W, Ferreira CEL (2011) Abundance and diversity patterns of the sessile macrobenthic community associated with environmental gradients in Vitoria Harbor, southeastern Brazil. Zoologia 28:641–652. https://doi.org/10.1590/s1984-46702011000500012

Zanella TV (2010) Água de lastro: um problema ambiental global. Curitiba: Juruá

Chapter 7
Research on Rocky Shores in Brazil: Advances and Contributions to International Fora

André Pardal, Stuart R. Jenkins, Sérgio A. Navarrete, Kringpaka Wangkulangkul, and Ronaldo A. Christofoletti

Abstract In this chapter, we provide an overview of how rocky shore research in Brazil contributes to international fora. Here, we bring our perspectives on the advance of Brazil-based science on the theme and its impact on global discussions. Collaborating with international co-authors, we draw parallels with studies in other regions that, like Brazil, have begun to contribute to rocky shore discussions at a later stage. Particularly, this literature has expanded our knowledge beyond temperate regions in the northern hemisphere. We argue that, building upon the groundwork laid by pioneering marine scientists, Brazil-based rocky shore research has now established itself within the international community, making notable contributions to various discussions on coastal ecology. This is evident in the insights it provides into our remarkable biodiversity, ecological patterns and processes, and human pressures exerted upon these environments.

A. Pardal (✉) · R. A. Christofoletti
Instituto do Mar, Universidade Federal de São Paulo (IMar/UNIFESP), Santos, SP, Brazil
e-mail: andre.pardal@unifesp.br

S. R. Jenkins
School of Ocean Sciences, Bangor University, Menai Bridge, Anglesey, UK

S. A. Navarrete
Estación Costera de Investigaciones Marinas and Center for Advanced Studies in Ecology and Biodiversity (CASEB), Departamento de Ecología, Facultad de Ciencias Biológicas, Pontificia Universidad Católica de Chile, Santiago, Chile

K. Wangkulangkul
Division of Biological Science, Faculty of Science, Prince of Songkla University, Songkhla, Thailand

R. Coutinho, R. A. Christofoletti (eds.), *Brazilian Rocky Shores*, Brazilian Marine Biodiversity, https://doi.org/10.1007/978-3-031-67206-4_7

Introduction

The rocky shore environment and its diverse biota have long stirred human curiosity and sparked scientific thinking. Studies on such habitats were crucial for developing and enhancing ecological theory, and more widely, biology. At a time when science was not organised in any sophisticated manner into disciplines, naturalists dedicated themselves to studying the amazing forms and interactions of living organisms on rocky shores. For instance, Charles Darwin's fascination with barnacles resulted in advances in the comprehension of morphology, palaeontology, systematics, and reproductive biology (Newman 1993; Castilla 2009). Such knowledge was also a source of ideas for his most monumental work: *On the Origin of Species* (Deutsch 2010)—and perhaps the real reason for delaying its publication (van Wyhe 2007). Some decades later, another British Charles, Charles Elton, published one of the foundational textbooks of ecology. In *Animal Ecology*, Elton (1927) considered the zonation of rocky intertidal organisms when discussing species distribution and succession. In the second half of the twentieth century, manipulative experiments on rocky shores, in addition to large-scale surveys, were instrumental in advancing community ecology. These studies crystallised the role of species interactions, abiotic disturbance, and the supply of resources modulating community structure and dynamics (see syntheses in Menge 1992, 2000a, b; Power 1992; Underwood 2000; Estes et al. 2016; Hawkins et al. 2020).

The foundational work on rocky shore ecology was developed in the UK and west coast of the USA (e.g. Stephenson and Stephenson 1949; Crisp and Southward 1958; Lewis 1964; Connell 1961a, b; Paine 1966; Dayton 1971, 1975; Menge and Sutherland 1976; see Hawkins et al. 2020 for a review). The largely descriptive approaches of the 1940s and the 1950s gave way to the experimental hypothesis-testing approaches by the late 1960s. As a result of international education and collaboration, and migration of professionals, a second wave of rocky shore studies emerged with impetus in the early to mid-1970s in the southern hemisphere, led by countries such as South Africa and Australia (e.g. Branch 1971, 1975; Underwood 1974, 1978). From the early 1980s on, active groups of rocky shore researchers were established in several institutions and countries around the world (Hawkins et al. 2020), and in South America, the studies from the cold productive waters of Chile started to become known (see reviews in Castilla and Paine 1987; Santelices 1990; Fernández et al. 2000; Thiel et al. 2007; Navarrete et al. 2010). In Brazil, early studies on rocky shores date back to at least the 1940s (Ghilardi et al. 2008; Moreno and Rocha 2012; Coutinho et al. 2016). However, national scientific production in this field became more significant and consistent only after the late 1990s (Coutinho et al. 2016; Mantelatto et al. 2018) (Fig. 7.1), following the consolidation of research groups. Currently, rocky shore research in Brazil, devoted to the study of both intertidal and shallow subtidal habitats and their associated biodiversity, is a flourishing field comprising many investigators and institutions all over the country, and covering a multitude of research topics.

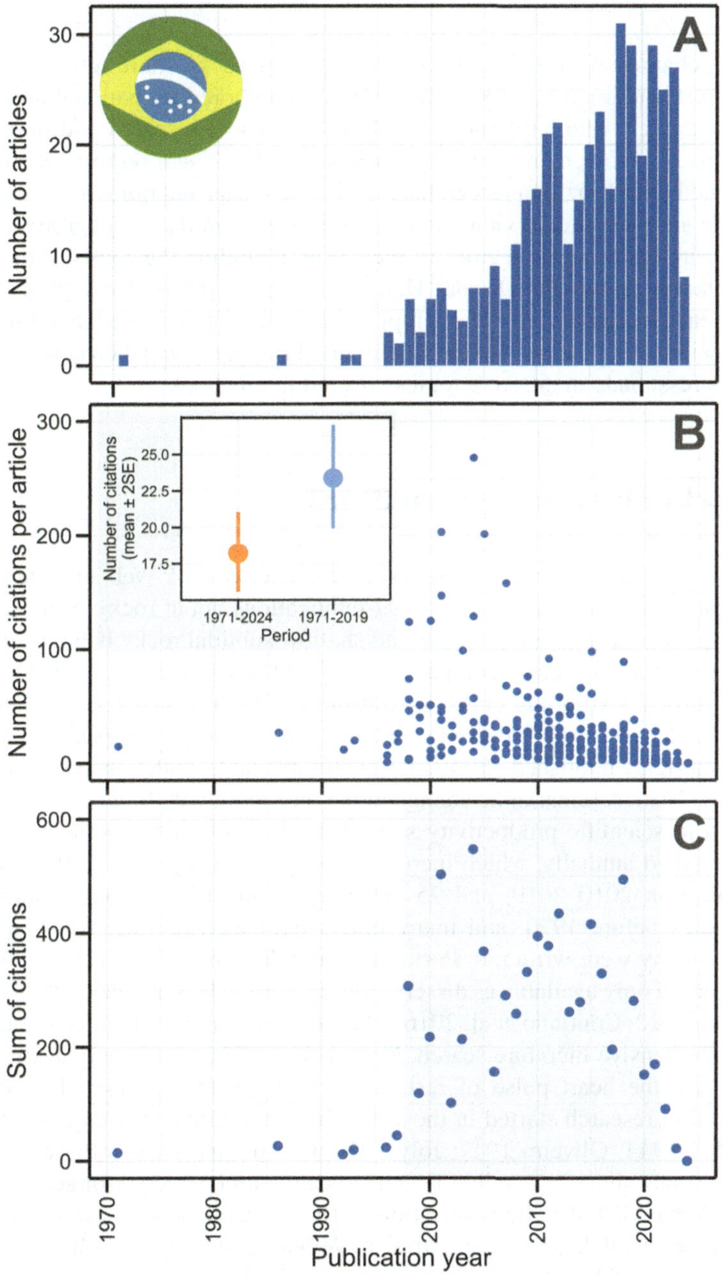

Fig. 7.1 (**a**) Number of articles retrieved in our searches in the database of Web of Science Core Collection in April/2024 about rocky shore research in Brazil over the years (see Annexe 7.1 for search details). (**b**) Number of citations of every article published between 1971 and 2024 (*n* = 407). In detail, the mean (±2SE) number of citations per article for 1971–2024 (*n* = 407) and 1971–2019 (*n* = 299) periods. (**c**) Sum of overall citations grouped per publication year of the articles (e.g. all articles published in 1998, 1999, and 2000 amounted to, respectively, 307, 119, and 218 citations). The grey rectangle highlights the 1998–2021 period

In this chapter, we provide an overview of how rocky shore research in Brazil contributes to international fora. This text is co-authored by national and international researchers who have participated in rocky shore research at different times and extents of involvement over the past few decades. The foreign co-authors have collaborated with Brazilian researchers in diverse forms and times and were invited to provide an international view on the matter, based on their background in rocky shore research at different regions of the world, including the coasts of the Pacific, North Atlantic, and Indian Oceans. Here we bring our perspectives on how Brazil-based science has advanced and attempt to assess its impacts on global discussions. We start with a brief analysis of the Brazilian bibliography and then comment on the main progress made in different topics covered in this book.

The Rocky Shore Research in Brazil

Based on two non-extensive searches in the database of Web of Science Core Collection in April/2024, we found 407 publications about rocky shore studies in Brazil. Our search covered intertidal and shallow subtidal rocky habitats, including sandstone formations, and encompassed both benthic and reef-associated biota (e.g. fish, sea turtles, swimming crabs) (see Annexe 7.1). The first record found dated back to 1971 and it was followed by almost two decades of apparent stasis, with only two articles published in internationally available outlets between 1971 and 1989 (Fig. 7.1a). A remarkable change took place in the 1990s, when a strong positive trend in scientific productivity started. In the 1990–1999 period, 1.6 articles were published annually, which increased to 7.7 articles/year in 2000–2009, 20.4 articles/year in 2010–2019, and 25 articles/year in 2020–2023. Naturally, there were studies before 1970, and many others published between 1971 and 1999. However, many were written in Portuguese and disseminated in national and local journals or are only available as dissertations or theses (Ghilardi et al. 2008; Moreno and Rocha 2012; Coutinho et al. 2016). Still, others were not captured in our admittedly non-extensive literature search. Nonetheless, if internationally available publications are the heart pulse of research activity in the country, the trend of a strengthening research started in the 1990s from the pioneering early studies (e.g. Rawitscher 1944; Oliveira 1947; Joly 1957; Coelho and Ramos-Porto 1980) suggests that rocky shore research in Brazil is healthy and getting stronger (Fig. 7.1a).

As of April/2024, the 407 publications retrieved in our searches were cited a total of 7640 times in all databases according to Web of Science Core Collection metrics, resulting in a global mean of 18.4 citations/article. When considering only research published up to 2019 ($n = 299$), the citation performance increases to 23.5 citations/article (Fig. 7.1b). The sum of citations grouped per publication year of the articles reveals a substantial increase in citation metrics of the Brazilian rocky shore bibliography after the late 1990s (Fig. 7.1c). The literature published between 1998 and 2021 amounted to an average of 300 citations (min-max: 102–547 citations) (Fig. 7.1c). Thus, it seems clear that rocky shore research in Brazil is not only increasing in productivity but that also the impact of the Brazilian research has increased significantly over the past three decades.

Building Foundations: Biodiversity, Patterns, and Physical Conditions

Early Brazilian rocky shore studies focused on describing biodiversity and patterns of distribution of intertidal organisms (Ghilardi et al. 2008; Moreno and Rocha 2012; Coutinho et al. 2016), thus following the natural course of scientific enquiry, building foundations to allow later work to focus on understanding ecological processes (Underwood et al. 2000). Such literature paved the way for more complex studies that started to address and disentangle the underlying processes and mechanisms that maintain biodiversity and variability of Brazilian tropical and subtropical systems. This was particularly important because classical rocky shore research has been grounded on studies carried out in northern hemisphere temperate areas, and the regional expansion of rocky shore studies in the 1970s was also concentrated on the highly productive temperate ecosystems. Brazilian studies revealed, for instance, interesting and somewhat divergent patterns in comparison to temperate rocky habitats, such as the high diversity of intertidal macroalgae (e.g. Oliveira 1947; Joly 1957; Ugadim 1973; Guimaraens and Coutinho 1996) and their unique associated fauna (e.g. Tararam 1977), as well as the elevated number of mobile invertebrate species (Vermeij and Porter 1971). These studies also introduced to the world the Brazilian "players" in the rocky shore arena advancing taxonomy and biogeography in addition to being fundamental for the identification of exotic species arriving over the last few decades (see Chap. 6).

Descriptive studies in Brazil, influenced by the long-established notion that rocky shore biota is strongly affected by environmental gradients (Elton 1927; Stephenson and Stephenson 1949; Crisp and Southward 1958), have quantified species distribution and abundance across shore levels (i.e. zonation) (Ghilardi et al. 2008). Such a focus is not surprising since Brazilian shores have among the clearest and visually most striking zonation patterns resulting from aggregated vertical distribution of dominant species across the intertidal zone. Moreover, such obvious differences in the distribution of dominant species in vertical zones were long believed, and demonstrated, to reflect boundaries of tolerance of abiotic stress and species interactions (Coutinho 1995). Investigations in Brazil generally have found patterns which converge with classical literature on intertidal zonation with abundant macroalgae low on the shore, and benthic invertebrates (i.e. mussels, barnacles) dominating the mid- and high-shore levels (e.g. Oliveira 1947; Coelho and Ramos-Porto 1980; Coutinho 1995; Masi and Zalmon 2008; Masi et al. 2009; Christofoletti et al. 2011a; Martinez 2012; Zamprogno et al. 2012; Correia et al. 2018; see also Chap. 4).

The degree of exposure to wave action is another important abiotic influence in rocky intertidal habitats (McQuaid and Branch 1985; Jenkins et al. 2008) also explored in Brazil. While the coastline of S-SE Brazil, where most rocky shores occur (Chap. 1), experiences lower levels of wave action than some other parts of the world (e.g. the Northeast Atlantic; Jenkins et al. 2008), the complex coastline does create strong gradients of wave exposure. Thus, addressing the influence of wave exposure on Brazilian shores provided tests of the broad role of this structuring factor. In general, comparisons between exposed and sheltered locations, or measurements along gradients of wave exposure, demonstrated patterns in distribution, abundance, diversity, biomass, and morphology of species which are in agreement with established patterns worldwide. For example, towards sites with increased wave action, intertidal zonation range enlarges (Masi and Zalmon 2008), sessile invertebrates become more abundant, bare rock becomes less common and community composition changes (Széchy and Paula 2000; Christofoletti et al. 2011a; Pardal et al. 2023), larval recruitment increases (Dias et al. 2018), and benthic species exhibit morphological alterations which decrease dislodgement risk (Tanaka et al. 2002; Vieira and Bueno 2019).

Diving beneath the tides, studies of shallow subtidal rocky habitats have contributed equally to highlighting Brazilian biodiversity, understanding patterns of distribution and identifying their environmental drivers. Compared to the intertidal zone, the subtidal domain holds a far greater diversity of species in much less accessible habitats. Assessing and identifying such tremendous biodiversity has required great efforts from taxonomists and ecologists—and for many of them, diving skills. Among other taxa, these studies have revealed a huge number of species of macroalgae, cnidarians, poriferans, ascidians, polychaetes, bryozoans, echinoderms, and reef-associated fish (e.g. Eston et al. 1986; Yoneshigue and Valentin 1988; Castro et al. 1995, 1999; Ferreira et al. 2001, 2015; Oigman-Pszczol et al. 2004; Bouzon et al. 2012; Bumbeer et al. 2016; Dias et al. 2013; Gibran and Moura 2012). All this endeavour proved to be truly important for detecting and studying bioinvasion, since a significant part of marine non-indigenous species in Brazil are associated with subtidal habitats (Neves et al. 2007; Ignacio et al. 2010; Oricchio et al. 2019; Teixeira and Creed 2020).

Regarding the identification of patterns of distribution and environmental drivers, studies in shallow subtidal rocky environments have highlighted the primary importance of depth in shaping benthic assemblages and associated mobile fauna. Similar to shore levels above water, depth governs significant gradients in other relevant variables affecting species, such as light, temperature, hydrodynamics, and salinity (Mantelatto et al. 2020a, 2022). Consequently, benthic assemblages are typically dominated by macroalgae in shallower areas, with an increasing coverage of sessile invertebrates towards greater depths (e.g. Eston et al. 1986; Oigman-Pszczol et al. 2004; Oigman-Pszczol et al. 2004; Magalhães et al. 2015). Likewise, abundance, diversity, biomass, and composition of sea urchins and reef fish have often been demonstrated to vary along depth gradients (e.g. Ferreira et al. 2001, 2015; Pereira-Filho et al. 2011; Gibran and Moura 2012; Cordeiro et al. 2014, 2016; Labbé-Bellas et al. 2016).

Moving on from Patterns to Processes

The use of manipulative field experiments is a growing practice in Brazilian rocky shore research, which has resulted in significant advances in understanding community dynamics (e.g. Sauer Machado et al. 1996). A better understanding of local causal-effect relationships and, to some extent, of ecosystem dynamics has frequently been achieved by experimental manipulations, such as predator exclusion experiments. The principle is simple: if consumption is an important regulatory process, substantial changes should occur after removing predators (Paine 1966). Compared to temperate systems, Brazilian rocky shores lack many of the benthic consumers previously shown to occupy keystone roles such as the famous predatory seastar *Pisaster ochraceus* (Paine 1966), the predatory snail *Concholepas concholepas* (Castilla 1999), or the grazing limpet *Patella vulgata* (Jenkins et al. 2008). Investigations in Brazil on the role of lesser known, often small-bodied, consumers in structuring intertidal communities have been one way that Brazil-based research has contributed towards global fora.

In Brazilian subtropical rocky shores, the grapsid crab *Pachygrapsus transversus* is a common and abundant mobile omnivore. In contrast to carnivory and herbivory, the role of omnivory in rocky intertidal habitats was less documented and not completely understood. Such observations led Christofoletti et al. (2010) to test the effect of diverse consumers, through exclusion experiments, on the community structure in the low intertidal zone of sheltered rocky shores in Ubatuba (São Paulo state, SE Brazil). The manipulations were designed to disentangle the effects of *Pachygrapsus* foraging from other taxa such as fish, large crabs, whelks, and limpets. While the exclusion of these consumers did not affect assemblage composition, the exclusion of the crab *P. transversus* resulted in substantial changes in the abundance of functional groups. The dominant groups covering the substratum changed from macroalgal species to an assemblage of suspensivore invertebrates (ascidians, sponges, and bivalves) (Christofoletti et al. 2010).

Experiments at Ubatuba also explored the effects of grazers on biofilm. Christofoletti et al. (2011b) did not find the expected changes in biofilm biomass in the mid- and high-shore levels following the exclusion of limpets and littorinids—which in many systems are important benthic herbivores capable of controlling algal biomass (e.g. Mak and Williams 1999; Aguilera 2011). However, the authors noticed that the exclusion treatments they had implemented were unable to exclude the isopod *Ligia exotica*, an abundant fast-moving omnivorous consumer. After further observations and experiments, Christofoletti et al. (2011b) concluded that these isopods were actually the main consumers controlling biofilm biomass in the mid- and high-shore levels of sheltered rocky shores, probably masking the effects of slow-moving grazers and abiotic influences.

Christofoletti et al. (2010, 2011b) showed that strong top-down control of subtropical intertidal communities is not exerted by the iconic, large, and usually strictly carnivore species, but by small benthic fast-moving omnivorous consumers. A picture started to cautiously surface from this still insufficient number of studies, which suggests that in low-productivity intertidal shores, high biomass of high-trophic level species cannot be attained by slow-moving carnivores. Instead, highly mobile omnivores, capable of consuming plant and animal tissue and moving over comparatively large areas, can attain population biomasses high enough to exert local top-down control. To what extent *Pachygrapsus* or *Ligia* can be considered keystone species depends largely on how strictly this concept is defined (Menge et al. 1994; Power et al. 1996). The results of Christofoletti et al. show the switch from the typically high *per capita* to high *per population* interaction strength (Navarrete and Menge 1996) of consumers in the intertidal community.

Further intertidal experiments carried out along environmental gradients have elucidated how physical conditions can modulate species interactions (Flores et al. 2015; Pardal et al. 2022), providing other home-brewed tests of classical ecological models (Menge 1992, 2000a). Flores et al. (2015) investigated the effects of wave action and grazing in regulating intertidal macroalgal biomass along shores in the São Sebastião Channel (SE Brazil). This study revealed that herbivore control (by limpets and littorinids) was restricted to sheltered sites where macroalgal growth was slower. Interestingly, Flores et al. (2015) proposed that variation in wave intensity not only establishes a gradient of potential disturbance or amelioration of desiccation stress, but it can also be a modulator of nutrient input. They suggested that in such oligotrophic waters, waves can locally increase the flux of nutrients that goes through intertidally attached fronds, thus favouring algal growth rate and yearly persistence.

Pardal et al. (2022) conducted local and large-scale experiments to investigate how environmental conditions affect predator-prey interactions. The first experiment, at multiple sites in Ubatuba, showed that the consumption rate and interaction strength of the predatory whelk *Stramonita brasiliensis* on the barnacle *Tetracita stalactifera* depended on wave exposure and prey density. These results confirmed that increasing environmental harshness can decrease the importance of predation (Menge and Sutherland 1976), but the magnitude of change depends on density-dependent effects (Berlow et al. 1999). Furthermore, a large-scale experiment carried out between Guarujá/SP and Arraial do Cabo/RJ did not reveal spatial patterns in the same whelk-barnacle interaction, nor relationships to nearshore chlorophyll-a concentration (Chl-a; as a proxy for food for the barnacles) or the minor variation in sea temperature during the study. Instead, Pardal et al. (2022) noted patterns in the size of consumed barnacles along the Chl-a gradient suggesting changes in whelk's food choice related to prey quality and size.

Further down on the shore, research in the subtidal have also been advancing our understanding of how predation affects community dynamics. By applying sophisticated strategies to manipulate predator and prey traits and predation context, these studies have revealed the complex ways in which environmental heterogeneity can alter the effects of consumers on benthic biota. For instance, in shallow subtidal rocky reefs from Arraial do Cabo (Rio de Janeiro state, SE Brazil), Ferreira et al. (1998) showed that the territorial damselfish *Stegastes fuscus* strongly regulates algal assemblage within its territories, keeping algal assemblage in an early succession stage and preventing dominance by the superior competitor algae. In the same region, Cordeiro et al. (2020) have shown that sea urchins at shallow depths, but not fish, controlled macroalgal cover. In deeper waters, no effect of either sea urchin or fish grazing was detected. These patterns revealed an ecosystem dynamic similar to temperate systems, although the biota is more similar to that of the tropics. Other studies on the effects of predation (mostly) by fish on subtidal fouling communities developed on experimental plates have established the importance of the timing of predation (Vieira et al. 2012), successional stage and recruitment rate of prey (Vieira et al. 2016), predator size (Oricchio et al. 2016), defensive traits of prey (Dias et al. 2020), and identity of founders (Vieira et al. 2021).

Impactful studies aiming to comprehend or infer bottom-up mechanisms of community regulation have also been made on Brazilian shores. A number of them have applied correlational analyses to link variability in resource availability (i.e. phytoplankton and biofilm biomass, nutrients) to larval properties (Kasten and Flores 2013; Kasten et al. 2019; Leal et al. 2021), larval settlement/recruitment (Mazzuco et al. 2015, 2018, 2019; Barbosa et al. 2016), or population metrics of adult suspensivores (Pardal et al. 2021, 2023; Cordeiro et al. 2024). Others have used isotopic analyses (e.g. Vinagre et al. 2018; Andrades et al. 2024). This literature brings interesting and novel understanding to global fora showing how benthic species couple to pelagic processes in a generally nutrient-depleted region (see Chap. 3). This is a different environmental context to temperate or high-productivity areas where most work has been done (e.g. Bustamante et al. 1995; Menge et al. 1997; Nielsen and Navarrete 2004). Brazilian research also highlights the differences between the influence of strong upwelling regimes occurring on western continental margins and the intermittent and more localised upwelling regimes on the Brazilian coast (mostly reported in the Cabo Frio region) (see Chap. 1). The work confirming the role of meteorological cold fronts as regulatory mechanisms of benthic-pelagic coupling in SE Brazil brought particularly original and exciting knowledge to the field (Mazzuco et al. 2015, 2018, 2019). Similar dynamics appear to hold true for Thailand tropical system, which, like Brazil, has recently become a focal point of investigation. Wangkulangkul et al. (2023), for instance, showed the role of monsoon-influenced oceanographic processes in affecting recruitment patterns of intertidal species.

One of the most classical and interesting demonstrations of bottom-up mechanisms of community regulation, almost omnipresent in ecological textbooks, is the influence of seabird guano on rocky shore communities. Guano has been shown to boost nutrient levels in the surrounding seawater, increasing rates of benthic primary production and, thereby, the growth and abundance of consumers (Bosman and Hockey 1986; Bosman et al. 1986). Curiously, long before Bosman's work, Charles Darwin himself got intrigued by a weird white substance covering the rocks when he spotted and landed in the São Pedro and São Paulo Archipelago (SPSPA) in 1832 during the H.M.S. Beagle trip (Darwin 1845). Almost two centuries after Darwin's intrigue, Andrades et al. (2024) traced guano incorporation in intertidal fish and crabs living in SPSPA tide pools, through stable isotope analysis, confirming the role of this source of nutrients in subsidising the local food web.

Seeking to See the Big Picture: Scaling-Up in Space and Time

The question of scale is central to ecology. Every observation is made over a particular spatiotemporal extent at a given resolution; hence, the conclusions derived from them are dependent on such context (Wiens 1989; Levin 1992). The development of community regulation theory, for instance, has stressed the importance of situating patterns and processes within the scales at which they occur—and at which they were measured. Initially, generalisations were made based on mechanisms investigated at local scales, ignoring influences emerging at broader extents (Lewin 1986; Underwood and Fairweather 1989; Menge 1992). Many of these generalisations were later disputed and refuted.

Brazilian literature has followed similar steps to global science regarding the spatial scaling-up of ecological studies. Early research was mainly focused on quantifying variation in populations, assemblages, and ecological processes in a single or a few sites. As discussed previously in this chapter, those efforts advanced (and keep advancing) our understanding of population and community dynamics and the influence of abiotic disturbance and species interactions (see Chap. 4). However, on an extensive and heterogeneous coast like that of Brazil, it has only been through macroecological studies that it has been possible to start seeing the big picture. As detailed in Chap. 2, attempts to determine biogeographical patterns of the rocky shore biota along the extensive Brazilian coast have long been around. However, studies addressing ecological questions *per se* (see Kent 2005) are mostly newer as discussed below.

It is interesting to note that the shift from local to large-scale approaches has occurred at different times for the studies on intertidal and subtidal rocky habitats depending on the target taxa. Macroecological investigations focusing on reef fish—including those associated with coral reefs—showed up earlier in Brazilian rocky shore research (e.g. Floeter et al. 2001, 2004, 2005; Ferreira et al. 2004). These unique studies were influential in revealing geographical variation in patterns of reef fish groups and assemblages (e.g. biomass, abundance, composition, and diversity) along the Brazilian coast—and sometimes beyond—and identifying potential environmental drivers. For example, among other variables, sea temperature was consistently related to the patterns uncovered in these studies. The impactful research tradition on macroecology of reef fish remains active and fruitful (e.g. Bender et al. 2013; Longo et al. 2014, 2019; Morais et al. 2017; Andrades et al. 2018; Pinheiro et al. 2018; see Chap. 2).

The benthic biota associated with rocky shores, on the other hand, had to wait a little longer before being subjected to macroecological assessments. Aued et al. (2018) studied subtidal benthic communities on shallow reefs (rocky and biogenic) spanning 27° of latitude (0°–27°S) along the Brazilian coast. This study showed that these habitats were dominated by algal turfs and frondose macroalgae, which drove community patterns regardless of reef type and geographic region. Interestingly, benthic diversity was lower in the tropics, contrary to the general pattern of latitudinal diversity gradient (Aued et al. 2018). More recently, Santana et al. (2023)

provided a similar macroecological study to that of Aued et al. (2018), identifying turbidity as the primary factor explaining the difference between assemblages dominated by turf algae and fleshy macroalgae. Turf and fleshy forms were prevalent, respectively, in turbid and clear-water environments (Santana et al. 2023).

Regarding intertidal benthic communities, sampling done in Brazil contributed to studies aiming at understanding coarse ecological patterns at the global (Cruz-Motta et al. 2010) and continental levels (Cruz-Motta et al. 2020). Other newer studies have advanced our understanding of how intertidal populations and communities vary in response to environmental variables emerging at multiple spatial scales. Following a similar approach, Pardal et al. (2021, 2023) and Cordeiro et al. (2024) investigated spatial variability in populations and communities over a wide range of scales (from meters to over 500 km) and their relationships with natural and anthropogenic variables along the coasts of São Paulo and Rio de Janeiro states. In general, these studies, in agreement with described patterns worldwide (e.g. Underwood and Chapman 1996; Fraschetti et al. 2005; Valdivia et al. 2014; Wangkulangkul and Promdam 2018; Carneiro et al. 2021), found that most of the variability in biota occurs at the smallest spatial scales assessed. Where variation occurred at larger scales, on the other hand, there were consistent influences of variables related to oceanic forcing (i.e. wave exposure, ocean temperature, Chl-a). In particular, the association of biotic responses to ocean temperature highlighted the long recognised (see Chaps. 2 and 4) Cabo Frio upwelling as an important influence on rocky shore components. Furthermore, shifts in population patterns and community structure near urbanised bays emphasised the potential impact of human activities on coastal benthic habitats. The decadal losses in canopy-forming algae and proliferation of turf forms in rocky shores demonstrated by Gorman et al. (2020), and also evidenced by Pardal et al. (2023), reinforce this relationship (see also De Paula et al. 2020).

In addition to evaluating the influence of abiotic factors on rocky intertidal species, Cordeiro et al. (2024) tested correlative models of top-down and bottom-up regulation. These analyses considered relationships between consumers and the characteristics of their food (abundance/biomass and body size). Cordeiro et al. (2024) found no or little evidence of bottom-up and top-down influences on intertidal populations. Such results agree with the manipulations carried out by Pardal et al. (2022) mentioned before, which also did not find large-scale patterns in a whelk-barnacle interaction. These findings indicated that over broad scales in SE Brazil, environmental factors dictate patterns of intertidal communities, while trophic interactions can be more important locally as demonstrated by experimental manipulations. Nevertheless, further studies addressing the role of predator-prey interactions and environmental variation over larger scales are needed in Brazil once those are still scarce.

In ecology, time is as important as space. Different environmental drivers vary at particular temporal scales, so understanding their influence depends on *when* observations are made and often for *how long* (Underwood 2000; Jenkins and Uyà 2016). For instance, tides vary over a matter of hours, wave action may intensify or lessen within days, and solar radiation changes within months. Other factors vary more markedly at larger temporal scales: among seasons (e.g. temperature, rain) or years (e.g. ENSO events). In Brazil, studies dedicated to understanding larval recruitment have particularly provided neat examples of the importance of identifying temporal scales of variation.

Studies in SE Brazil (Mazzuco et al. 2015, 2019) found that the strength of the relationship between recruitment and oceanic-climatic forcing (e.g. Chl-a, sea temperature, wind patterns, cold fronts) depended on the temporal scale evaluated—days, weeks, or months—with taxa-specific trends. Mazzuco et al. (2015), for instance, showed that barnacle recruitment was positively correlated with wind direction at the scale of weeks, which was an opposite trend compared to that observed in monthly and daily analyses. Negative correlations of barnacle recruitment with cold fronts and Chl-a were also detected at the scale of months, but not weeks. On the other hand, weekly trends in mussel recruitment were positively related to wind speed, while daily trends negatively covaried with sea temperature. The same negative influence of cold fronts on barnacle recruitment at the month scale (Mazzuco et al. 2015) was also found for assemblage recruitment of benthic organisms associated with macroalgae in sandstone reefs in Espírito Santo state (Mazzuco et al. 2019).

Mazzuco and Bernardino (2022) evaluated whether spatiotemporal dynamics in larval recruitment of invertebrates associated with macroalgae could be predicted by variability in seascapes, a detailed classification of integrated pelagic conditions (i.e. Chl-a, sea temperature, and salinity) based on satellite products (distributed by NOAA/CoastWatch). Analyses over 2 years revealed that monthly and seasonal variations in assemblage recruitment and composition correlated with seascape classifications. Recruitment rates were strongly influenced by subtropical pelagic conditions and were reduced in warm waters with high-nutrient contents and phytoplankton blooms. Mazzuco and Bernardino (2022) provided novel insights into the application of an integrative approach in benthic ecology, evidencing how larval supply could be impacted by ocean changes.

The comprehension of benthic-pelagic coupling in Brazil contributes to a broader discussion concerning the significance of these processes in low-productivity subtropical and tropical regions, which are often underrepresented in global research. Efforts to integrate these areas into the framework of rocky shore ecology gained momentum in the 1990s, propelled by notable studies conducted in Hong Kong (see Williams et al. 2019 for a review). More recently, investigations conducted across Southeast Asia have also provided valuable insights. For instance, studies in the intertidal zone of the Thai-Malay peninsula have revealed that the recruitment of the barnacle *Chthamalus malayensis* occurs throughout the year with a few peaks (Lee et al. 2006; Wangkulangkul et al. 2023). This pattern appears to mirror the recruitment behaviour of its Brazilian counterpart, *Chthamalus bisinuatus*, which exhibits sporadic recruitment throughout the year, with a pronounced peak during the summer months along the São Paulo coast (Barbosa et al. 2016; Dias et al. 2018). Conversely, research in the Cabo Frio region has indicated a higher incidence of barnacle recruitment during autumn and early winter, likely attributed to the influence of upwelling (Skinner and Coutinho 2002). Recruitment of another key space occupier in many shores in SE Asia, the oyster *Saccostrea* spp. is generally poor without distinct peaks with population maintenance relying on sporadic episodes of robust recruitment occurring every few years (Wangkulangkul et al. 2023). Similarly, transient dominance of the Brazilian mussel *Mytilaster solisianus* in São Sebastião/SP was attributed to poor recruitment across years (Petersen et al. 1986). Studies on recruitment are key for understanding the maintenance of adult populations and ultimately ecosystem functioning (Underwood and Fairweather 1989; Menge 2000a, b). Moreover, comparing the role of recruitment in regulating intertidal communities between different tropical-subtropical regions on different continents is essential for reaching a better understanding of rocky shore ecology at the global scale.

In a rare long-term study, Puga et al. (2019) investigated temporal variation—at the scales of fortnight, month, season, and year—in the cover of intertidal benthic organisms on a rocky shore in Guanabara Bay/RJ for 7 years. The abundance of the dominant species (oysters and barnacles) varied over the years but not seasons, with the opposite patterns for the algae *Ulva* spp. Puga et al. also demonstrate that the influence of environmental drivers on species abundance varied at different temporal scales, with species-specific trends depending on weekly, fortnightly, or monthly variation in abiotic variables (i.e. seawater and air temperature, tides, and pollution input). This study highlights the importance of biodiversity assessments over longer durations to better understand ecosystem dynamics (see Navarrete et al. 2010), as advocated by Jenkins and Uyà (2016). It is worth noting that this work is an output of a project within the Brazilian Long-Term Ecological Research Program (PELD) supported by the National Council for Scientific and Technological Development (CNPq). PELD has been central for advancing Brazilian research towards scaling-up studies in time (see Cordeiro et al. 2022) and, consequently, addressing global questions.

In summary, the bibliography visited in this section has enhanced our comprehension of spatial and temporal variation in patterns and processes on rocky shores integrating benthic ecology and coastal oceanography. In addition to their value in contributing to macroecology in an understudied region and globally, these investigations should be (and have been) useful for detecting and forecasting changes resulting from unfolding human impacts. As such, they also contribute to managerial actions towards marine spatial planning and conservation.

Biting the Bullet: Addressing Human Pressures on Rocky Shores

Characterised by diverse socio-cultural contexts, the extensive Brazilian coastal zone experiences multiple uses which exert pressures varying in type and intensity across regions. As discussed in Chap. 6, research on human pressures on rocky shores remains largely concentrated in a limited number of locations, highlighting significant scientific gaps in various ecological and environmental topics. However, Brazilian research has provided valuable insights to international scientific discussions regarding challenges in developing countries, along with their socio-economic and ecological impacts (e.g. Ferreira et al. 2009; Martinez et al. 2022; Eggertsen et al. 2024).

Brazilian shores experience intense artisanal fisheries, which has impacted several reef fish populations (Floeter et al. 2006; Eggertsen et al. 2024). Yet, the impacts on other organisms, especially invertebrates, remain poorly documented due to the absence of comprehensive fishery monitoring programs (Neto et al. 2021; Eggertsen et al. 2024). Similar to countries like Thailand, Chile, Portugal, and South Africa, harvesting on rocky shores involves the collection of invertebrates such as snails, oysters, mussels, octopuses, crabs, and sea urchins for consumption (e.g. Kyle et al. 1997; Moreno 2001; Rius and Cabral 2004; Lage and Jablonski 2008). In Brazil, studies indicate that a significant number of species are illegally harvested, many of which are primarily destined for the aquarium trade (see Chap. 6). However, data on quantities and spatiotemporal patterns of harvesting of these organisms remains unknown, and there is an urgent need for comprehensive monitoring and management to sustain these tropical and subtropical ecosystems.

Pollution is another major issue in Brazil. In particular, inadequate sewage treatment infrastructure poses widespread risks to human communities and ecosystems (Martinez et al. 2022). Despite being an all-pervasive threat, the impacts of sewage discharge on rocky shores are still poorly understood. To partially address this matter, some research has used proxies of sewage contamination alongside other urbanisation metrics (e.g. Martins et al. 2012; de Vasconcelos et al. 2019). Although these efforts do not allow determination of causative relationships, they have evidenced changes in macroalgae assemblages towards the dominance of opportunistic groups in more urbanised areas (and under increased impact of sewage discharge). Such a pattern corroborates the role of nutrient pollution in shifting marine communities in favour of *r*-selected species (Johnston et al. 2022). Moreover, monitoring sewage contamination is necessary to better comprehend its impacts, which is often a challenging task. Gorman et al. (2017) offer a neat example of how using isotopic analysis of intertidal species can help with that. The study revealed that algae effectively tracked short-term nitrogen pollution in coastal waters, while hermit crabs indicated longer-term nitrogen inputs. This research highlighted the value of using multiple taxa to monitor spatial and temporal patterns of nitrogen pollution.

Brazilian research has also advanced our understanding of how pollution impacts the functioning of marine organisms and ecosystems, addressing a critical gap in global research (Johnston et al. 2015). Martinez et al. (2019) showed that, with increased metal concentrations along the Guarujá coast (São Paulo state, SE Brazil) as a consequence of urbanisation, the mussel *Mytilaster solisianus* increased filtration and biodeposition rates. These results indicated that urban-associated contamination stresses the bivalves, potentially disrupting benthic-pelagic coupling on rocky shores. Many other Brazilian studies have explored the structural and physiological responses of rocky shore organisms to diverse pollutants. Such literature has shown alterations in biota ranging from the individual to the assemblage level (see review in Martinez et al. 2022). These have been important for documenting the impacts of pollution all around the country and placing the Brazilian reality within the international context.

Recent international discussions have increasingly focused on the impacts of coastal infrastructure on marine environments (Firth et al. 2024). As in other countries, like Chile, Thailand, and Malaysia (e.g. Aguilera 2018; Lai et al. 2018; Wangkulangkul and Promdam 2018; Chee et al. 2021), research in this area is developing in Brazil with interesting contributions. Pardal-Souza et al. (2017), for instance, have demonstrated that shading cast by coastal structures strongly impacts rocky intertidal biota. Field observations, confirmed by manipulative experiments, revealed that shading dramatically reduces macroalgae biomass and shifts recruitment and community patterns, favouring sessile suspensivores over macroalgae. This work was conducted in the context of the recent planning for expansion of the Port of São Sebastião (São Paulo state), within a large multidisciplinary project (Turra et al. 2017). The Araçá Project was an ecosystem-based research program developed at Araçá Bay, an adjacent area that would be the most impacted by eventual port expansion. As with many other outputs from the Araçá Project (see the full issue in Ocean & Coastal Management volume 164, introduced by Amaral et al. 2018), Pardal-Souza et al. (2017) provided critical information for supporting local management.

In an effort to assess what we already know about the ecological impacts of artificial constructions in Brazil, Martinez et al. (under review) provided a comprehensive overview of the scientific knowledge on the theme. The review showed that national research has focused on artificial reefs and their influence on fish populations and assemblages. This work also revealed an elevated number of invasive exotic species found in association with artificial substrates, evidencing their known role in facilitating bioinvasion (e.g. Neves et al. 2007; Bulleri and Chapman 2010; Bumbeer and Rocha 2012; Oricchio et al. 2019). In another recent work, Pardal et al. (2024) provide the first large-scale mapping of hard artificial structures in Brazil. After assessing the extent of artificial structures on the entire coast of the São Paulo state, the authors quantified a total linear extent of 244 km of artificial structures, with 63% of them occurring in the most urbanised region (Baixada Santista). Breakwalls, jetties, and wharves represented almost 80% of all infrastructure mapped. Given the role of artificial hard substrates in facilitating bioinvasion, which is a source of impact on rocky shores (Chap. 6), the results of Pardal et al. present key information for managerial strategies towards conserving biodiversity and sustaining ecosystem services in such habitats.

Research on marine bioinvasion *per se* is very active in Brazil, resulting in robust scientific advances. Both in intertidal and sublittoral environments, many investigations have long strived to detect non-indigenous species and study their expansion in distribution and impacts on native biodiversity (reviews in Ferreira et al. 2009; Teixeira and Creed 2020; and see also Chap. 6). Of particular concern has been the rapid expansion of the sun-coral *Tubastraea* spp. along the Brazilian coast (see Creed et al. 2017), with studies identifying potential vectors such as marine litter (Mantelatto et al. 2020b) and oil platforms as ecological stepping stones (Coelho et al. 2022). Furthermore, the impacts of this invasive species in our ecosystems have been demonstrated, including decreasing recruitment, altering the abundance of native species, and shifting assemblage structure (e.g. Miranda et al. 2018; Silva et al. 2019; Tanasovici et al. 2020). Such extensive literature, together with that focused on other non-indigenous species (see Chap. 6), have addressed key questions in international fora, strengthening our knowledge arsenal to manage this complex and widespread environmental problem.

Although research on climate change in Brazil is in the early stages, it has already enriched global understanding of its impacts in tropical and subtropical regions (Coutinho et al. 2016). Key findings include potential distribution shrinkage of zoanthids linked to changes in salinity, pH, and temperature (Durante et al. 2018), as well as the decline in macroalgal populations in S-SE Brazil associated with rising ocean temperatures over the past 50 years (Gorman et al. 2020). Additionally, some laboratory experiments have indicated that physiological responses of rocky shore consumers are altered under scenarios of climate change. Madeira et al. (2018) showed that the whelk *Stramonita brasiliensis* has high thermal tolerance and acclimation ability in the short, but not in the long term. Thus, exposure to chronic warming resulted in negative effects on growth and weight gain. In another study, using a bifactorial design considering seawater temperature and pH, Rich et al. (2018) showed that under a scenario of ocean acidification (i.e. ambient temperature, high-CO_2) metabolic rates and algal consumption of the sea urchin *Echinometra lucunter* increased. If these laboratory studies are extrapolated to natural conditions, ocean warming and acidification must have significant disruptive consequences for predator-prey interactions in Brazilian rocky shores. On the other hand, Leal et al. (2020) demonstrated that the barnacle *Chthamalus bisinuatus*, which occupies the high-shore level, may be resilient to elevated substrate temperature, providing insights into species-specific adaptations.

Concluding Remarks

Building upon the groundwork laid by pioneering marine biologists and oceanographers in the country, Brazil-based rocky shore research has now firmly established itself within the international community, making notable contributions to various discussions on coastal ecology. Throughout this chapter, we explored a bibliography that showcases advances in the field and their contributions to discussions beyond national context and applications. We discussed the evolution of our research practices, which increasingly incorporate experimental hypothesis-testing approaches and address environmental complexities, while also expanding the spatial and temporal dimensions of investigations. Of course, such comprehensive approaches do not overshadow studies employing descriptive methods, which remain valuable for addressing specific inquiries. The insights gained from research on Brazilian ecosystems, spanning rocky shores and other habitats along the coast, have also played a significant role in informing conservation efforts and promoting sustainable management practices. Finally, it is worth acknowledging that while our selection of articles in this chapter was not exhaustive, the omission of certain works does not diminish their merit or contributions in any way.

Acknowledgements We are thankful to Dra Aline S. Martinez for her valuable comments on the early versions of this chapter. This work was supported by Conselho Nacional de Desenvolvimento Científico e Tecnológico (CNPq-AP: 385628/2023-4; RAC: 317015/2023-0) and São Paulo Research Foundation (FAPESP-RAC: 2016/24551-4).

Annexe 7.1 Details in Literature Search

Published papers about rocky shore studies in Brazil were gathered using the search engine Web of Science Core Collection (until April 21, 2024) with the following keywords: **TOPIC** (rocky shore* OR rocky intertidal OR rocky subtidal OR rocky reef* OR hard bottom* OR rocky bottom* OR subtidal rocky reef* OR shallow reef* OR beachrock reef* OR sandstone reef*) **AND TOPIC** (benthic community* OR distribution* OR abundance*) **AND TOPIC** (Brazil* OR Southwest* Atlantic) **AND ADDRESS** (Brazil). This search resulted in *416* papers. We read the title of all papers, and when these were not informative enough to decide on their inclusion based on our selection criteria, we also read their abstracts. After this procedure, we excluded 124 papers that (i) were not about rocky shores; (ii) were not done in Brazil; (iii) or were not related to biodiversity. This first search resulted in **292** papers.

Another wider search was done on April 22, 2024, to complement the first one, using the following keywords: **TOPIC** (rocky shore* OR rocky intertidal OR rocky reef* OR hard bottom*) **AND TOPIC** (Brazil), resulting in *456* papers. After filtering the results by reading the titles and when necessary the abstracts based on our inclusion criteria, we kept *249* results. We then merged the search dataset (292 + 249 = 541) and removed *134* duplicates. Therefore, the two searches resulted in **407** unique papers considered in this chapter.

References

Aguilera MA (2011) The functional roles of herbivores in the rocky intertidal systems in Chile: a review of food preferences and consumptive effects. Rev Chil Hist Nat 84(2):241–261

Aguilera MA (2018) Artificial defences in coastal marine ecosystems in Chile: opportunities for spatial planning to mitigate habitat loss and alteration of the marine community structure. Ecol Eng 120:601–610

Amaral ACZ, Ciotti AM, Fonseca G (2018) Biodiversity and functioning of a subtropical coastal ecosystem: subsidies for integrated management. Ocean Coast Manag 164:1–3

Andrades R, Machado FS, Reis-Filho JA, Macieira RM, Giarrizzo T (2018) Intertidal biogeographic subprovinces: local and regional factors shaping fish assemblages. Front Mar Sci 5:412

Andrades R, Cardozo-Ferreira GC, Benevides LJ, Pimentel CR, Mancini PL, Ferreira CEL, Giarrizzo T, Joyeux JC, Macieira RM (2024) Seabird guano reshapes intertidal reef food web in an isolated oceanic islet. Coral Reefs 43:347–355

Aued AW, Smith F, Quimbayo JP, Candido DV, Longo GO, Ferreira CEL, Witman JD, Floeter SR, Segal B (2018) Large-scale patterns of benthic marine communities in the Brazilian Province. PLoS One 13(6):e0198452

Barbosa ACC, Gomes CC, Pereira GC, Bueno M, Flores AAV (2016) Local biological drivers, not remote forcing, predict settlement rate to a subtropical barnacle population. Mar Ecol Prog Ser 543:201–208

Bender MG, Pie MR, Rezende EL, Mouillot D, Floeter SR (2013) Biogeographic, historical and environmental influences on the taxonomic and functional structure of Atlantic reef fish assemblages. Glob Ecol Biogeogr 22(11):1173–1182

Berlow EL, Navarrete SA, Briggs CJ, Power ME, Menge BA (1999) Quantifying variation in the strengths of species interactions. Ecology 80(7):2206–2224

Bosman AL, Hockey PAR (1986) Seabird guano as a determinant of rocky intertidal community structure. Mar Ecol Prog Ser 18:247–257

Bosman AL, Du Toit JT, Hockey PAR, Branch GM (1986) A field experiment demonstrating the influence of seabird guano on intertidal primary production. Estuar Coast Shelf Sci 23(3):283–294

Bouzon JL, Brandini FP, Rocha RM (2012) Biodiversity of sessile fauna on rocky shores of coastal islands in Santa Catarina, Southern Brazil. Mar Sci 2(5):39–47

Branch GM (1971) The ecology of *Patella linnaeus* from the cape Peninsula, South Africa I. Zonation, movements and feeding. Afr Zool 6(1):1–38

Branch GM (1975) Ecology of *Patella* species from the Cape Peninsula, South Africa. IV. Desiccation. Mar Biol 32:179–188

Bulleri F, Chapman MG (2010) The introduction of coastal infrastructure as a driver of change in marine environments. J Appl Ecol 47(1):26–35

Bumbeer JA, Rocha RMD (2012) Detection of introduced sessile species on the near shore continental shelf in southern Brazil. Zoologia 29:126–134

Bumbeer J, Cattani AP, Chierigatti NB, Rocha RM (2016) Biodiversity of benthic macroinvertebrates on hard substrates in the Currais Marine Protected Area, in southern Brazil. Biota Neotrop 16:e20160246

Bustamante RH, Branch GM, Eekhout S (1995) Maintenance of an exceptional intertidal grazer biomass in South Africa: subsidy by subtidal kelps. Ecology 76(7):2314–2329

Carneiro IM, Bertocci I, Paiva PC, Széchy MTM (2021) Natural and anthropogenic factors as possible drivers of variability in rocky shore assemblages at multiple spatial scales. Estuar Coast Shelf Sci 262:107577

Castilla JC (1999) Coastal marine communities: trends and perspectives from human-exclusion experiments. Trends Ecol Evol 14(7):280–283

Castilla JC (2009) Darwin taxonomist: barnacles and shell burrowing barnacles. Rev Chil Hist Nat 82:477–483

Castilla JC, Paine RT (1987) Predation and community organization on Eastern Pacific, temperate zone, rocky intertidal shores. Rev Chil Hist Nat 60:131–151

Castro CB, Echeverría CA, Pires DO, Mascarenhas BJA, Freitas SG (1995) Distribuição de Cnidaria e Echinodermata no infralitoral de costões rochosos de Arraial do Cabo, Rio de Janeiro, Brasil. Rev Bras Biol 55:471–480

Castro CB, Echeverría CA, Pires DO, Fonseca CG (1999) Distribuição de bentos (Cnidaria e Echinodermata) em costões rochosos da Baía da Ilha Grande, Rio de Janeiro, Brasil. Oecol Bras 7:179–193

Chee SY, Yee JC, Cheah CB, Evans AJ, Firth LB, Hawkins SJ, Strain EMA (2021) Habitat complexity affects the structure but not the diversity of sessile communities on tropical coastal infrastructure. Front Ecol Evol 9:673227

Christofoletti RA, Murakami VA, Oliveira DN, Barreto RE, Flores AAV (2010) Foraging by the omnivorous crab *Pachygrapsus transversus* affects the structure of assemblages on sub-tropical rocky shores. Mar Ecol Prog Ser 420:125–134

Christofoletti RA, Takahashi CK, Oliveira DN, Flores AAV (2011a) Abundance of sedentary consumers and sessile organisms along the wave exposure gradient of subtropical rocky shores of the south-west Atlantic. J Mar Biol Assoc UK 91(5):961–967

Christofoletti RA, Almeida TV, Ciotti AM (2011b) Environmental and grazing influence on spatial variability of intertidal biofilm on subtropical rocky shores. Mar Ecol Prog Ser 424:15–23

Coelho PA, Ramos-Porto M (1980) Bentos litorâneos do nordeste oriental do Brasil II. Povoamentos de substratos duros. Bol Inst Oceanogr 29:133–134

Coelho SCC, Gherardi DFM, Gouveia MB, Kitahara MV (2022) Western boundary currents drive sun-coral (*Tubastraea* spp.) coastal invasion from oil platforms. Sci Rep 12(1):5286

Connell JH (1961a) The influence of interspecific competition and other factors on the distribution of the barnacle *Chthamalus stellatus*. Ecology 42(4):710–723

Connell JH (1961b) Effects of competition, predation by *Thais lapillus*, and other factors on natural populations of the barnacle *Balanus balanoides*. Ecol Monogr 31(1):61–104

Cordeiro CAMM, Harborne AR, Ferreira CEL (2014) Patterns of distribution and composition of sea urchin assemblages on Brazilian subtropical rocky reefs. Mar Biol 161:2221–2232

Cordeiro CAMM, Mendes TC, Harborne AR, Ferreira CEL (2016) Spatial distribution of nominally herbivorous fishes across environmental gradients on Brazilian rocky reefs. J Fish Biol 89(1):939–958

Cordeiro CAMM, Harborne AR, Ferreira CEL (2020) The biophysical controls of macroalgal growth on subtropical reefs. Front Mar Sci 7:488

Creed JC, Fenner D, Sammarco P, Cairns S, Capel K, Junqueira AOR, Cruz I, Miranda RJ, Carlos-Júnior L, Mantelatto MC, Oigman-Pszczol S (2017) The invasion of the azooxanthellate coral Tubastraea (*Scleractinia: Dendrophylliidae*) throughout the world: history, pathways and vectors. Biol Invasions 19:283–305. https://doi.org/10.1007/s10530-016-1279-y

Cordeiro CAMM, Aued AW, Barros F, Bastos AC, Bender M, Mendes TC, Creed JC, Cruz IC, Dias MS, Fernandes LD, Coutinho R (2022) Long-term monitoring projects of Brazilian marine and coastal ecosystems. PeerJ 10:e14313

Cordeiro CAMM, Pardal A, Noya LG, Ciotti AM, Jenkins SR, Burrows MT, Williams GA, Christofoletti RA (2024) Environmental factors have stronger effects than biotic processes in patterns of intertidal populations along the southeast coast of Brazil. Mar Environ Res 200:106646. https://doi.org/10.1016/j.marenvres.2024.106646

Correia JRM, Oliveira WD, Pereira PS, de Camargo JMR, de Araújo ME (2018) Substrate zonation as a function of reef morphology: a case study in Carneiros Beach, Pernambuco, Brazil. J Coast Res 81:1–9

Coutinho R (1995) Avaliação crítica das causas da zonação dos organismos bentônicos em costões rochosos. Oecol Bras 1:259–271

Coutinho R, Yaginuma LE, Siviero F et al (2016) Studies on benthic communities of rocky shores on the Brazilian coast and climate change monitoring: status of knowledge and challenges. Braz J Oceanogr 64(sp2):27–36

Crisp DJ, Southward AJ (1958) The distribution of intertidal organisms along the coasts of the English Channel. J Mar Biol Assoc UK 37(1):157–203

Cruz-Motta JJ, Miloslavich P, Palomo G et al (2010) Patterns of spatial variation of assemblages associated with intertidal rocky shores: a global perspective. PLoS One 5(12):e14354

Cruz-Motta JJ, Miloslavich P, Guerra-Castro E et al (2020) Latitudinal patterns of species diversity on South American rocky shores: local processes lead to contrasting trends in regional and local species diversity. J Biogeogr 47(9):1966–1979

Darwin C (1845) Journal of researches into the natural history and geology of the countries visited during the voyage of H.M.S. Beagle round the world: under the command of Capt. Fitz Roy, R.N. John Murray, London

Dayton PK (1971) Competition, disturbance, and community organization: the provision and subsequent utilization of space in a rocky intertidal community. Ecol Monogr 41(4):351–389

Dayton PK (1975) Experimental evaluation of ecological dominance in a rocky intertidal algal community. Ecol Monogr 45(2):137–159

De Paula JC, Lopes-Filho EAP, de Carvalho WF, Coração ACS, Yoneshigue-Valentin Y (2020) Long-term changes in macroalgae assemblages reveal a gradual biodiversity loss over the last 200 years in the hypereutrophic Guanabara Bay. Mar Environ Res 162:105153

De Vasconcelos ERTPP, Vasconcelos JB, Reis TNV, Cocentino ALM, Mallea AJA, Martins GM, Neto AI, Fujii MT (2019) Macroalgal responses to coastal urbanization: relative abundance of indicator species. J Appl Phycol 31:893–903

Deutsch J (2010) Darwin and barnacles. C R Biol 333(2):99–106

Dias GM, Rocha RM, Lotufo TMC, Kremer LP (2013) Fifty years of ascidian biodiversity research in São Sebastião, Brazil. J Mar Biol Assoc UK 93(1):273–282

Dias GM, Christofoletti RA, Kitazawa K, Jenkins SR (2018) Environmental heterogeneity at small spatial scales affects population and community dynamics on intertidal rocky shores of a threatened bay system. Ocean Coast Manag 164:52–59

Dias GM, Vieira EA, Pestana L, Marques AC, Karythis S, Jenkins SR, Griffith K (2020) Calcareous defence structures of prey mediate the effects of predation and biotic resistance towards the tropics. Divers Distrib 26(9):1198–1210

Durante LM, Cruz ICS, Lotufo TMC (2018) The effect of climate change on the distribution of a tropical zoanthid (*Palythoa caribaeorum*) and its ecological implications. PeerJ 6:4777

Eggertsen L, Luza AL, Cordeiro CAMM et al (2024) Complexities of reef fisheries in Brazil: a retrospective and functional approach. Rev Fish Biol Fish 34:511–538

Elton CS (1927) Animal ecology. The MacMillian Company, New York

Estes JA, Dayton PK, Kareiva P, Levin SA, Lubchenco J, Menge BA, Palumbi SR, Power ME, Terborgh J (2016) A keystone ecologist: Robert Treat Paine, 1933–2016. Ecology 97(11):2905–2909

Eston VR, Migotto AE, Oliveira Filho EC, Rodrigues SA, de Freitas JC (1986) Vertical distribution of benthic marine organisms on rocky coasts of the Fernando de Noronha Archipelago (Brazil). Bol Inst Oceanogr 34:37–53

Fernández M, Jaramillo E, Marquet PA, Moreno CA, Navarrete SA, Ojeda FP, Valdovinos C, Vásquez J (2000) Diversity, ecology and biogeography of nearshore benthic ecosystems: an overview and needs for conservation. Rev Chil Hist Nat 73:629–662

Ferreira CEL, Gonçalves JEA, Coutinho R, Peret AC (1998) Herbivory by the dusky damselfish *Stegastes fuscus* (Cuvier, 1830) in a tropical rocky shore: effects on the benthic community. J Exp Mar Biol Ecol 229:241–264

Ferreira CEL, Gonçalves JEA, Coutinho R (2001) Community structure of fishes and habitat complexity on a tropical rocky shore. Environ Biol Fish 61:353–369

Ferreira CEL, Floeter SR, Gasparini JL, Ferreira BP, Joyeux JC (2004) Trophic structure patterns of Brazilian reef fishes: a latitudinal comparison. J Biogeogr 31:1093–1106

Ferreira CEL, de Oliveira Ribeiro Junqueira A, Villac MC, Lopes RM (2009) Marine bioinvasions in the Brazilian coast: brief report on history of events, vectors, ecology, impacts and management of non-indigenous species. In: Rilov G, Crooks JA (eds) Biological invasions in marine ecosystems. Ecological studies, vol 204. Springer, Berlin, Heidelberg

Ferreira CM, Coni EOC, Medeiros DV, Sampaio CL, Reis-Filho JA, Barros F, Loiola M, Nunes JACC (2015) Community structure of shallow rocky shore fish in a tropical bay of the southwestern Atlantic. Braz J Oceanogr 63:379–396

Firth LB, Bone J, Bartholomew A et al (2024) Coastal greening of grey infrastructure: an update on the state-of-the-art. ICE Proc Maritime Eng 177:35–67

Floeter SR, Guimarães RZP, Rocha LA, Ferreira CEL, Rangel CA, Gasparini JL (2001) Geographic variation in reef-fish assemblages along the Brazilian coast. Glob Ecol Biogeogr 10:423–431

Floeter SR, Ferreira CEL, Dominici-Arosemena A, Zalmon IR (2004) Latitudinal gradients in Atlantic reef fish communities: trophic structure and spatial use patterns. J Fish Biol 64:1680–1699

Floeter SR, Behrens MD, Ferreira CEL, Paddack MJ, Horn MH (2005) Geographical gradients of marine herbivorous fishes: patterns and processes. Mar Biol 147:1435–1447

Floeter SR, Halpern BS, Ferreira CEL (2006) Effects of fishing and protection on Brazilian reef fishes. Biol Conserv 128(3):391–402

Flores AAV, Christofoletti RA, Peres ALF, Ciotti AM, Navarrete SA (2015) Interactive effects of grazing and environmental stress on macroalgal biomass in subtropical rocky shores: modulation of bottom-up inputs by wave action. J Exp Mar Biol Ecol 463:39–48

Fraschetti S, Terlizzi A, Benedetti-Cecchi L (2005) Patterns of distribution of marine assemblages from rocky shores: evidence of relevant scales of variation. Mar Ecol Prog Ser 296:13–29

Ghilardi NP, Pereira Filho GH, Berchez F (2008) Current knowledge status of the ecology of hard bottom benthic communities in Brazil and the need for new approaches. Oecol Bras 12(2):197–205

Gibran FZ, Moura RLD (2012) The structure of rocky reef fish assemblages across a nearshore to coastal islands' gradient in Southeastern Brazil. Neotrop Ichthyol 10:369–382

Gorman D, Turra A, Connolly RM, Olds AD, Schlacher TA (2017) Monitoring nitrogen pollution in seasonally-pulsed coastal waters requires judicious choice of indicator species. Mar Pollut Bull 122:149–155

Gorman D, Horta P, Flores AAV et al (2020) Decadal losses of canopy-forming algae along the warm temperate coastline of Brazil. Glob Chang Biol 26:1446–1457

Guimaraens MA, Coutinho R (1996) Spatial and temporal variation of benthic marine algae at the Cabo Frio upwelling region, Rio de Janeiro, Brazil. Aquat Bot 52(4):283–299

Hawkins SJ, Pack KE, Hyder K, Benedetti-Cecchi L, Jenkins SR (2020) Rocky shores as tractable test systems for experimental ecology. J Mar Biol Assoc UK 100(7):1017–1041

Ignacio BL, Julio LM, Junqueira AO, Ferreira-Silva MA (2010) Bioinvasion in a Brazilian bay: filling gaps in the knowledge of southwestern Atlantic biota. PLoS One 5(9):e13065

Jenkins SR, Uyà M (2016) Temporal scale of field experiments in benthic ecology. Mar Ecol Prog Ser 547:273–286

Jenkins SR, Moore P, Burrows MT, Garbary DJ, Hawkins SJ, Ingólfsson A, Sebens KP, Snelgrove PV, Wethey DS, Woodin SA (2008) Comparative ecology of North Atlantic shores: do differences in players matter for process? Ecology 89(sp11):S3–S23

Johnston EL, Mayer-Pinto M, Crowe TP (2015) Chemical contaminant effects on marine ecosystem functioning. J Appl Ecol 52:140–149

Johnston EL, Clark GF, Bruno JF (2022) The speeding up of marine ecosystems. Clim Change Ecol 3:100055

Joly AB (1957) Contribuição ao conhecimento da flora ficológica marinha da Baía de Santos e arredores. Bol Assoc Geogr Bras 217:7–27

Kasten P, Flores AAV (2013) Disruption of endogenous tidal rhythms of larval release linked to food supply and heat stress in an intertidal barnacle. Mar Ecol Prog Ser 472:185–198

Kasten P, Tremblay R, Flores AAV (2019) Mismatched seasonal patterns of larval production and quality in subtropical barnacle populations along a coastal trophic gradient. Estuar Coast Shelf Sci 224:43–50

Kent M (2005) Biogeography and macroecology. Prog Phys Geogr 29(2):256–264

Kyle R, Pearson B, Fielding PJ, Robertson WD, Birnie SL (1997) Subsistence shellfish harvesting in the Maputaland marine reserve in northern KwaZulu-Natal, South Africa: rocky shore organisms. Biol Conserv 82(2):183–192

Labbé-Bellas R, Cordeiro CAMM, Floeter SR, Segal B (2016) Sea urchin abundance and habitat relationships in different Brazilian reef types. Reg Stud Mar Sci 8:33–40

Lage H, Jablonski S (2008) Mussel Perna perna extraction and commercialization in Guanabara Bay, Brazil. Rev Atlântica 30(2):161–169

Lai S, Loke LH, Bouma TJ, Todd PA (2018) Biodiversity surveys and stable isotope analyses reveal key differences in intertidal assemblages between tropical seawalls and rocky shores. Mar Ecol Prog Ser 587:41–53

Leal I, Flores AAV, Archambault P, Collin R, Tremblay R (2020) Response of tropical and subtropical chthamalid barnacles to increasing substrate temperatures. J Exp Mar Biol Ecol 524:151281

Leal I, Bohn K, Hawkins SJ, Jenkins SR, Flores AAV, Tremblay R (2021) Lipid allocation in late-stage barnacle larvae from subtropical and temperate waters. Mar Ecol Prog Ser 661:147–161

Lee WJ, O'Riordan RM, Koh LL (2006) Spatial and temporal patterns in the recruitment of the intertidal barnacle *Chthamalus malayensis* Pilsbry (Crustacea: Cirripedia) on the equatorial shores of Peninsular Malaysia and Singapore. J Exp Mar Biol Ecol 333(2):296–305

Levin SA (1992) The problem of pattern and scale in ecology. Ecology 73:1943–1967

Lewin R (1986) Supply-side ecology. Science 234:25–27

Lewis JR (1964) The ecology of rocky shores. English Universities Press, London

Longo GO, Ferreira CEL, Floeter SR (2014) Herbivory drives large-scale spatial variation in reef fish trophic interactions. Ecol Evol 4(23):4553–4566

Longo GO, Hay ME, Ferreira CEL, Floeter SR (2019) Trophic interactions across 61 degrees of latitude in the Western Atlantic. Glob Ecol Biogeogr 28(2):107–117

Madeira C, Mendonça V, Flores AAV, Diniz MS, Vinagre C (2018) High thermal tolerance does not protect from chronic warming – a multiple end-point approach using a tropical gastropod, *Stramonita haemastoma*. Ecol Indic 91:626–635

Magalhães GM, Amado-Filho GM, Rosa MR, Moura RL, Brasileiro PS, Moraes FC, Francini-Filho RB, Pereira-Filho GH (2015) Changes in benthic communities along a 0–60 m depth gradient in the remote St. Peter and St. Paul Archipelago (Mid-Atlantic ridge, Brazil). Bull Mar Sci 91:377–396

Mak YM, Williams GA (1999) Littorinids control high intertidal biofilm abundance on tropical, Hong Kong rocky shores. J Exp Mar Biol Ecol 233(1):81–94

Mantelatto MC, Cruz ICS, Creed JC (2018) A review of the knowledge of shallow subtidal benthic communities in Brazil. Mar Environ Res 140:69–77

Mantelatto MC, Oliveira AES, Menegola C, Casares FA, Creed JC (2020a) Depth and grazing intensity are the main drivers of subtidal hardground benthic community structure on tropical south Atlantic reefs. Mar Ecol 41:e12586

Mantelatto MC, Póvoa AA, Skinner LF, de Araujo FV, Creed JC (2020b) Marine litter and wood debris as habitat and vector for the range expansion of invasive corals (*Tubastraea* spp.). Mar Pollut Bull 160:111659

Mantelatto MC, Carlos-Júnior LA, Côrrêa C, de Lima Cardoso CF, Creed JC (2022) Depth-related drivers of benthic community structure on shallow subtidal rocky reefs. Estuar Coast Shelf Sci 266:107743

Martinez AS (2012) Spatial distribution of the invasive bivalve *Isognomon bicolor* on rocky shores of Arvoredo Island (Santa Catarina, Brazil). J Mar Biol Assoc UK 92(3):495–503

Martinez AS, Mayer-Pinto M, Christofoletti RA (2019) Functional responses of filter feeders increase with elevated metal contamination: are these good or bad signs of environmental health? Mar Pollut Bull 149:110571

Martinez AS, Underwood T, Christofoletti RA, Pardal A, Fortuna MA, Marcelo-Silva J, Morais GC, Lana PC (2022) Reviewing the effects of contamination on the biota of Brazilian coastal ecosystems: scientific challenges for a developing country in a changing world. Sci Total Environ 803:150097

Martinez AS, Rocha RM, Christofoletti RA, Vieira LM, Pinheiro US, Haddad MA, Pinheiro IEG, Pellizari FM, Haraguchi L, Pardal A (under review) Ecological research on marine artificial structures in Brazil: Current knowledge and scientific guidelines to support sustainable management. (Under review in Ocean Coastal Management, June/2024)

Martins CD, Arantes N, Faveri C, Batista MB, Oliveira EC, Pagliosa PR, Fonseca AL, Nunes JMC, Chow F, Pereira SB, Horta PA (2012) The impact of coastal urbanization on the structure of phytobenthic communities in southern Brazil. Mar Pollut Bull 64(4):772–778

Masi BP, Zalmon IR (2008) Zonation of intertidal benthic communities on breakwaters of different hydrodynamics in the north coast of the state of Rio de Janeiro, Brazil. Rev Bras Zool 25:662–673

Masi BP, Macedo IM, Zalmon IR (2009) Benthic community zonation in a breakwater on the north coast of the state of Rio de Janeiro, Brazil. Braz Arch Biol Technol 52:637–646

Mazzuco ACA, Christofoletti RA, Pineda J, Starczak VR, Ciotti AM (2015) Temporal variation in intertidal community recruitment and its relationships to physical forcings, chlorophyll-a concentration and sea surface temperature. Mar Biol 162:1705–1725

Mazzuco ACA, Christofoletti RA, Coutinho R, Ciotti AM (2018) The influence of atmospheric cold fronts on larval supply and settlement of intertidal invertebrates: case studies in the Cabo Frio coastal upwelling system (SE Brazil). J Sea Res 137:47–56

Mazzuco ACA, Stelzer PS, Donadia G, Bernardino JV, Joyeux JC, Bernardino AF (2019) Lower diversity of recruits in coastal reef assemblages are associated with higher sea temperatures in the tropical South Atlantic. Mar Environ Res 148:87–98

Mazzuco ACA, Bernardino AF (2022) Reef larval recruitment in response to seascape dynamics in the SW Atlantic. Sci Rep 12:7750. https://doi.org/10.1038/s41598-022-11809-1

McQuaid CD, Branch GM (1985) Trophic structure of rocky intertidal communities: response to wave action and implications for energy flow. Mar Ecol Prog Ser 22(2):153–161

Menge BA (1992) Community regulation: under what conditions are bottom-up factors important on rocky shores? Ecology 73(3):755–765

Menge BA (2000a) Top-down and bottom-up community regulation in marine rocky intertidal habitats. J Exp Mar Biol Ecol 250(1–2):257–289

Menge BA (2000b) Recruitment vs. postrecruitment processes as determinants of barnacle population abundance. Ecol Monogr 70(2):265–288

Menge BA, Sutherland JP (1976) Species diversity gradients: synthesis of the roles of predation, competition, and temporal heterogeneity. Am Nat 110(973):351–369

Menge BA, Berlow EL, Blanchette CA, Navarrete SA, Yamada SB (1994) The keystone species concept: variation in interaction strength in a rocky intertidal habitat. Ecol Monogr 64(3):249–286

Menge BA, Daley BA, Wheeler PA, Dahlhoff E, Sanford E, Strub PT (1997) Benthic-pelagic links and rocky intertidal communities: bottom-up effects on top-down control? Proc Natl Acad Sci 94(26):14530–14535

Miranda RJ, Tagliafico A, Kelaher BP, Mariano-Neto E, Barros F (2018) Impact of invasive corals *Tubastrea* spp. on native coral recruitment. Mar Ecol Prog Ser 605:125–133

Morais RA, Ferreira CEL, Floeter SR (2017) Spatial patterns of fish standing biomass across Brazilian reefs. J Fish Biol 91(6):1642–1667

Moreno CA (2001) Community patterns generated by human harvesting on Chilean shores: a review. Aquat Conserv Mar Freshwat Ecosyst 11(1):19–30

Moreno TR, Rocha RM (2012) Ecologia de costões rochosos. Estud Biol, Ambiente Divers 34(83):191–201

Navarrete SA, Menge BA (1996) Keystone predation and interaction strength: interactive effects of predators on their main prey. Ecol Monogr 66(4):409–429

Navarrete SA, Gelcich S, Castilla JC (2010) Long-term monitoring of coastal ecosystems at Las Cruces, Chile: defining baselines to build ecological literacy in a world of change. Rev Chil Hist Nat 83:143–157

Neto JBG, Goyanna FAA, Feitosa CV, Soares MO (2021) A sleeping giant: the historically neglected Brazilian fishing sector. Ocean Coast Manag 209:105699

Neves CS, Rocha RM, Pitombo FB, Roper JJ (2007) Use of artificial substrate by introduced and cryptogenic marine species in Paranaguá Bay, southern Brazil. Biofouling 23:319–330

Newman WA (1993) Darwin and cirripedology. In: Truesdale FJ (ed) The history of carcinology. Crustacean issues 8. A. A. Balkema, Rotterdam, pp 349–434

Nielsen KJ, Navarrete SA (2004) Mesoscale regulation comes from the bottom-up: intertidal interactions between consumers and upwelling. Ecol Lett 7(1):31–41

Oliveira L (1947) Distribuição geográfica da fauna e flora da Baía da Guanabara. Mem Inst Oswaldo Cruz 45:709–735

Oigman-Pszczol SS, Figueiredo MAO, Creed JC (2004) Distribution of benthic communities on the tropical rocky subtidal of Armação dos Búzios, Southeastern Brazil. Mar Ecol 25(3):173–190

Oricchio FT, Flores AAV, Dias GM (2016) The importance of predation and predator size on the development and structure of a subtropical fouling community. Hydrobiologia 776:209–219

Oricchio FT, Marques AC, Hajdu E, Pitombo FB, Azevedo F, Passos FD, Vieira LM, Stampar SN, Rocha RM, Dias GM (2019) Exotic species dominate marinas between the two most populated regions in the southwestern Atlantic Ocean. Mar Pollut Bull 146:884–892

Paine RT (1966) Food web complexity and species diversity. Am Nat 100(910):65–75

Pardal A, Cordeiro CAMM, Ciotti AM, Jenkins SR, Gimenez L, Burrows MT, Christofoletti RA (2021) Influence of environmental variables over multiple spatial scales on the population structure of a key marine invertebrate. Mar Environ Res 170:105410

Pardal A, Jenkins SR, Christofoletti RA (2022) Local and large-scale spatial variation in a marine predator-prey interaction in the southwestern Atlantic. Oecologia 199(3):685–698

Pardal A, Martinez AS, Ciotti AM, Christofoletti RA, Cordeiro CAMM (2023) Macroecology of rocky intertidal benthic communities along the southwestern Atlantic: patterns of spatial variation and associations with natural and anthropogenic variables. Mar Environ Res 190:106099

Pardal A, Christofoletti RA, Martinez AS (2024) Urbanisation on the coastline of the most populous and developed state of Brazil: the extent of coastal hardening and occupations in low-elevation zones. Anthr Coasts 7:15. https://doi.org/10.1007/s44218-024-00048-8

Pardal-Souza AL, Dias GM, Jenkins SR, Ciotti AM, Christofoletti RA (2017) Shading impacts by coastal infrastructure on biological communities from subtropical rocky shores. J Appl Ecol 54(3):826–835

Pereira-Filho GH, Amado-Filho GM, Guimarães SM et al (2011) Reef fish and benthic assemblages of the Trindade and Martin Vaz island group, southwestern Atlantic. Braz J Oceanogr 59:201–212

Petersen JA, Sutherland JP, Ortega S (1986) Patch dynamics of mussel beds near São Sebastião (São Paulo), Brazil. Mar Biol 93:389–393

Pinheiro HT, Rocha LA, Macieira RM et al (2018) South-western Atlantic reef fishes: zoogeographical patterns and ecological drivers reveal a secondary biodiversity centre in the Atlantic Ocean. Divers Distrib 24:951–965

Power ME (1992) Top-down and bottom-up forces in food webs: do plants have primacy. Ecology 73(3):733–746

Power ME, Tilman D, Estes JA, Menge BA, Bond WJ, Mills LS, Daily G, Castilla JC, Lubchenco J, Paine RT (1996) Challenges in the quest for keystones: identifying keystone species is difficult—but essential to understanding how loss of species will affect ecosystems. Bioscience 46(8):609–620

Puga CA, Torres AS, Paiva PC, Yoneshigue-Valentin Y, Junqueira AO (2019) Multi-year changes of a benthic community in the mid-intertidal rocky shore of a eutrophic tropical bay (Guanabara Bay, RJ – Brazil). Estuar Coast Shelf Sci 226:106265

Rawitscher FK (1944) Algumas noções sobre a vegetação do litoral brasileiro. Bol Assoc Geogr Bras 5:13–28

Rich WA, Schubert N, Schläpfer N, Carvalho VF, Horta AC, Horta PA (2018) Physiological and biochemical responses of a coralline alga and a sea urchin to climate change: implications for herbivory. Mar Environ Res 142:100–107

Rius M, Cabral HH (2004) Human harvesting of *Mytilus galloprovincialis* Lamarck, 1819, on the central coast of Portugal. Sci Mar 68(4):545–551

Santana EF, Mies M, Longo GO, Menezes R, Aued AW, Luza AL, Bender MG, Segal B, Floeter SR, Francini-Filho RB (2023) Turbidity shapes shallow Southwestern Atlantic benthic reef communities. Mar Environ Res 183:105807

Santelices B (1990) Patterns of organizations of intertidal and shallow subtidal vegetation in wave exposed habitats of central Chile. Hydrobiologia 192:35–57

Sauer Machado KRS, Chapman ARO, Coutinho R (1996) Consumer species have limited and variable roles in community organization on a tropical intertidal shore. Mar Ecol Prog Ser 134:73–83

Silva R, Vinagre C, Kitahara MV, Acorsi IV, Mizrahi D, Flores AAV (2019) Sun coral invasion of shallow rocky reefs: effects on mobile invertebrate assemblages in Southeastern Brazil. Biol Invasions 21:1339–1350

Skinner LF, Coutinho R (2002) Preliminary results on settlement of the barnacles *Tetraclita stalactifera* and *Chthamalus bisinuatus* on a Brazilian tropical rocky shore under upwelling conditions. Invertebr Reprod Dev 41(1–3):151–156

Stephenson TA, Stephenson A (1949) The universal features of zonation between tide-marks on rocky coasts. J Ecol 37(2):289–305

Széchy MTM, Paula EJ (2000) Padrões estruturais quantitativos de bancos de *Sargassum* (Phaeophyta, Fucales) do litoral dos estados do Rio de Janeiro e São Paulo, Brasil. Rev Bras Bot 23:121–132

Tanaka MO, Duque-Estrada TE, Magalhães CA (2002) Dynamics of the acmaeid limpet *Collisella subrugosa* and vertical distribution of size and abundance along a wave exposure gradient. J Molluscan Stud 68(1):55–64

Tanasovici RM, Kitahara MV, Dias GM (2020) Invasive coral *Tubastraea* spp. population growth in artificial habitats and its consequences to the diversity of benthic organisms. Mar Biol 167(8):119

Tararam AS (1977) A fauna vágil de *Sargassum cymosum* C. Agardh, 1820 da Praia do Lamberto e Praia Grande, Ubatuba, São Paulo, com especial referência aos Gammaridea (Crustacea-Amphipoda). Master dissertation, Instituto Oceanográfico, Universidade de São Paulo, São Paulo

Teixeira LM, Creed JC (2020) A decade on: an updated assessment of the status of marine non-indigenous species in Brazil. Aquat Invasions 15(1):30–43

Thiel M, Macaya EC, Acuña E et al (2007) The Humboldt Current System of northern and central Chile. Oceanogr Mar Biol Annu Rev 45:95–344

Turra A, Amaral ACZ, Ciotti AM et al (2017) Environmental impact assessment under an ecosystem approach: the São Sebastião harbor expansion project. Ambient Soc 20(3):155–176

Ugadim Y (1973) Algas marinhas bentônicas do litoral sul do estado de São Paulo e do litoral do estado do Paraná. I Divisão Chlorophyta. Bol Bot Univ São Paulo 1:11–77

Underwood AJ (1974) The reproductive cycles and geographical distribution of some common eastern Australian prosobranchs (Mollusca: Gastropoda). Mar Freshw Res 25(1):63–88

Underwood AJ (1978) An experimental evaluation of competition between three species of intertidal prosobranch gastropods. Oecologia 33:185–202

Underwood AJ (2000) Experimental ecology of rocky intertidal habitats: what are we learning? J Exp Mar Biol Ecol 250(1–2):51–76

Underwood AJ, Chapman MG (1996) Scales of spatial patterns of distribution of intertidal invertebrates. Oecologia 107:212–224

Underwood AJ, Fairweather PG (1989) Supply-side ecology and benthic marine assemblages. Trends Ecol Evol 4:16–20

Underwood AJ, Chapman MG, Connell SD (2000) Observations in ecology: you can't make progress on processes without understanding the patterns. J Exp Mar Biol Ecol 250(1–2):97–115

Valdivia N, Díaz MJ, Holtheuer J, Garrido I, Huovinen P, Gomez I (2014) Up, down, and all around: scale-dependent spatial variation in rocky-shore communities of Fildes Peninsula, King George Island, Antarctica. PLoS One 9(6):e100714

van Wyhe J (2007) Mind the gap: did Darwin avoid publishing his theory for many years? Notes Rec R Soc 61(2):177–205

Vermeij GJ, Porter JW (1971) Some characteristics of the dominant intertidal molluscs from rocky shores in Pernambuco, Brazil. Bull Mar Sci 21(2):440–454

Vieira EA, Bueno M (2019) Small spatial scale effects of wave action exposure on morphological traits of the limpet *Lottia subrugosa*. J Mar Biol Assoc UK 99(6):1309–1315

Vieira EA, Duarte LFL, Dias GM (2012) How the timing of predation affects composition and diversity of species in a marine sessile community? J Exp Mar Biol Ecol 412:26–133

Vieira EA, Dias GM, Flores AAV (2016) Effects of predation depend on successional stage and recruitment rate in shallow benthic assemblages of the Southwestern Atlantic. Mar Biol 163:87
Vieira EA, Flores AAV, Dias GM (2021) Colonization history meets further niche processes: how the identity of founders modulates the way predation structure fouling communities. Oecologia 196(4):1167–1178
Vinagre C, Mendonça V, Flores AAV, Baeta A, Marques JC (2018) Complex food webs of tropical intertidal rocky shores (SE Brazil) – an isotopic perspective. Ecol Indic 95:485–491
Wangkulangkul K, Promdam R (2018) Spatial variability in the composition of macrofauna on intertidal rocky shores along the coast of the Andaman Sea and the Gulf of Thailand, Southern Thailand. Plankton Benthos Res 13(4):154–162
Wangkulangkul K, Meksuwan P, Stankovic M (2023) The role of monsoon-driven oceanographic variability in recruitment patterns of key intertidal space occupiers on tropical rocky shores on the west coast of Thailand, Indian Ocean. Hydrobiologia 850(16):3565–3579
Wiens JA (1989) Spatial scaling in ecology. Funct Ecol 3:385–397
Williams GA, Chan BKK, Dong YW (2019) Rocky shores of mainland China, Taiwan and Hong Kong: past, present and future. In: Hawkins SJ, Bohn K, Firth LB, Williams GA (eds) Interactions in the marine benthos: global patterns and processes, Systematics Association special volume series. Cambridge University Press, Cambridge, pp 360–390
Yoneshigue Y, Valentin JL (1988) Comunidades algais fotófilas de infralitoral de Cabo Frio, Rio de Janeiro, Brasil. Gayana Bot 45:61–75
Zamprogno GC, Fernandes FC, Fernandes LL (2012) Temporal and spatial variation of rocky shores intertidal benthic communities in Southeast Brazil. Iheringia Sér Zool 102:375–383

Chapter 8
Brazilian Rocky Shores: Perspectives for Future Research, Management, and Conservation

Ronaldo A. Christofoletti and Ricardo Coutinho

Abstract The aim of this book is to provide an overview of the current knowledge about rocky shores along the Brazilian coast to the national and international community. More than just a scientific review of published articles, this book encompasses the accumulated knowledge of researchers working in this environment. It aims to offer a synthesis that also benefits the training of human resources in undergraduate and graduate programs and other sectors of society interested in the topic. This book focused on consolidated substrates in the intertidal zones and shallow sublittoral areas, with coral reefs addressed in a separate volume. To achieve our goal, the current knowledge presented in Chaps. 1, 2, 3, 4, 5, 6 and 7 has been compiled by researchers from different fields of study in rocky shore research, from various institutions along the coast and at different stages of their careers, thus broadening the diversity of perspectives and expertise on the topics. In this chapter, we provide an integrative overview of Chaps. 1, 2, 3, 4, 5, 6, and 7 offering insight into the overall narrative of this book and addressing questions regarding knowledge gaps and future steps for research on Brazilian rocky shores. We conclude discussing the usage of this knowledge and the processes of management and conservation.

Advancements and Future Directions in Rocky Shore Research

This book serves as a comprehensive exploration of the dynamic ecosystems found along the Brazilian rocky shores, aiming to provide an invaluable resource for the global scientific community. Our journey through the chapters has offered a

R. A. Christofoletti (✉)
Instituto do Mar, Universidade Federal de São Paulo (IMar/UNIFESP), Santos, SP, Brazil
e-mail: christofoletti@unifesp.br

R. Coutinho
Instituto de Estudos do Mar Almirante Paulo Moreira, Departamento de Biotecnologia Marinha, Arraial do Cabo, RJ, Brazil

R. Coutinho, R. A. Christofoletti (eds.), *Brazilian Rocky Shores*, Brazilian Marine Biodiversity, https://doi.org/10.1007/978-3-031-67206-4_8

multifaceted examination of the intricate web of life that thrives within these environments. From the distribution and dynamics of intertidal communities to the cultural, economic, and environmental significance of rocky shores, each chapter has contributed to a deeper understanding of these ecosystems. Also, the synthesis of knowledge presented led us to identify some of the gaps that should be addressed in future research. Here, we present an overview of the main topics covered in each chapter followed by indication of gaps in knowledge based on the editors' experience. Therefore, we focus on the main topics and not specific research and citations, as the previous chapters have identified the proper bibliography extensively for each theme.

The narrative of this book began considering classical aspects of rocky shores, with Chaps. 1, 2, and 3 focusing on environmental features, biodiversity, biogeography, and benthic-pelagic coupling. These three chapters provide a holistic view of marine-coastal ecosystems, specifically focusing on Brazilian rocky shores, addressing aspects from their physical characteristics to biological interactions and human influences.

Calado et al. (2024) (Chapter 1) underscore the significance of rocky shores as biodiverse ecosystems shaped by intricate interactions of physical, chemical, and biological processes, notably emphasizing their role in species sustenance and evolutionary history. This chapter also outlines the susceptibility to erosion of rocky shores, both from natural forces like waves and winds and from bioerosion by marine organisms. Focusing on Southeast and South regions of Brazil, particularly Cabo Frio, where upwelling fosters marine life, Calado et al. (2024) discuss the impact of hydrodynamic processes, such as the Brazil Current, and wind-generated waves on coastal dynamics and marine ecosystems. Additionally, the authors highlight the importance of wave measurement and modeling for understanding oceanic climate and coastal interactions, emphasizing the need for a better comprehension to ensure the conservation and sustainable management of these vital coastal environments.

Cordeiro et al. (2024) (Chapter 2) provide an in-depth exploration of species richness, origin, and ecological processes in intertidal zones and rocky reefs across S-SE Brazil. The chapter highlights the higher species richness in the intertidal zone found in São Paulo state compared to macroalgae-rich coasts in Espírito Santo. In this chapter, Cordeiro et al. (2024) elucidate the phylogeographic patterns and distribution of intertidal organisms, shaped by geographical barriers and oceanographic influences. Additionally, it discusses the influence of environmental factors like ocean productivity and temperature on intertidal and reef communities, stressing potential challenges posed by anthropogenic pressures such as global warming. In addition, Fernandes and Da Gama (2024) (Chapter 3) delve into the intricate relationships between benthic and pelagic marine organisms, focusing on their roles in gamete dispersion, larval development, and reproductive success. Fernandes and Da Gama (2024) examine larval transport mechanisms, settlement dynamics, and the impact of anthropogenic factors like overfishing and pollution on benthic-pelagic coupling. By detailing these complex interactions, the chapter underscores the importance of understanding both natural processes and human influences in shaping marine ecosystem dynamics and biodiversity.

Concerning the topics covered in the first three chapters, we identified that studies about changes in environmental patterns due to climate change, such as shifts in waves and winds, are central for understanding future impacts on rocky shore dynamics. Also, it seems clear that there is a current knowledge gap about species distribution along the extensive Brazilian coast, with research being highly fragmented and concentrated on a few locations. Brazilian rocky shores are concentrated in the south and southeast coast, while sandstone reefs are predominant in the northeast. These habitats have similar ecosystem and seascape functions, thereby further research on integrating data from rocky shores and sandstone reefs can increase our comprehension of large-scale patterns. To address this, future studies should broaden their scope to encompass diverse taxonomic groups and incorporate both structural and functional response assessments of the biota and ecosystems over a range of spatial scales. By expanding research efforts to cover a wider geographical range and taxonomic diversity, scientists can fill the gaps in distribution knowledge and gain a more comprehensive understanding of rocky shore ecosystems. This approach will not only improve our knowledge of species distributions but also enhance our ability to assess and mitigate the impacts of environmental stressors such as climate change and bioinvasions on coastal biodiversity. Ultimately, bridging the gap in distribution knowledge will contribute to more effective management and conservation of rocky shore habitats, ensuring their long-term resilience and functionality in the face of global environmental change.

In sequence from the environmental and biodiversity knowledge, Zalmon et al. (2024) (Chapter 4), Ghilardi-Lopes et al. (2024) (Chapter 5), Martinez and Altvater (2024) (Chapter 6), and Pardal et al. (2024) (Chapter 7) offer a comprehensive insight into the complexity and significance of rocky coastal ecosystems in Brazil, addressing factors influencing the distribution and dynamics of populations and communities to the impacts of human activities and environmental changes, along with the contribution of this knowledge to international discussions.

Zalmon et al. (2024) (Chapter 4) provide an in-depth exploration of the complex dynamics within intertidal rocky shore communities, elucidating the interplay between biotic and abiotic factors that influence distribution patterns of biota. Highlighting the diverse array of influences such as emersion/desiccation, wave exposure, temperature, substrate orientation, and biological interactions such as competition, predation, and recruitment, the authors underscore the intricate mosaic of habitats within the intertidal zone. Zalmon et al. (2024) emphasize the dynamic balance between top-down and bottom-up controls, where consumers, resources, and environmental factors regulate predator and prey populations, shaping community structure and succession rates. Overall, the chapter illuminates the nuanced relationships between organisms and their environment in intertidal rocky shores, revealing the complex mechanisms that drive community dynamics in these ecosystems.

In this ecological context, future research efforts should address several key areas to advance our understanding of rocky shore ecosystems. One such area is the distribution and processes at large scales, which has been relatively understudied compared to other aspects. Additionally, there is a need to expand experimental manipulation in the field to better comprehend ecological interactions such as predation, herbivory, and facilitation, as well as the effects of stressor interactions on these processes. Moreover, studying eco-physiological aspects and the impacts of climate change on organisms inhabiting rocky shores is crucial for unraveling the intricate web of interactions within these ecosystems. By delving into the physiological traits of these organisms, future research can uncover how they adapt to changing environmental conditions, providing insights into their resilience or vulnerability in the face of climate change. Understanding these responses is paramount for predicting shifts in species distributions, community composition, and ecosystem functioning. Moreover, such studies shed light on the role of rocky shore organisms as indicators of environmental change, serving as early warning systems for broader ecological transformations. By elucidating these eco-physiological mechanisms, scientists can inform conservation strategies and management practices aimed at preserving the resilience and biodiversity of coastal ecosystems in the face of ongoing climate change challenges.

Ghilardi-Lopes et al. (2024) (Chapter 5) delve into the multifaceted socioeconomic significance of Brazilian rocky shores. First, their recreational and touristic allure is underscored, offering a range of activities from tide pool exploration to diving, contributing not only to local economies but also to individual well-being. Likewise, the pivotal role of rocky shore in environmental education and citizen science is emphasized, providing hands-on learning experiences and fostering an understanding of environmental changes. Culturally, these areas hold immense value, inspiring artistic expression and weaving into the historical tapestry of coastal communities through traditional fishing practices and indigenous cultures. Economically, they sustain livelihoods for artisanal fishing communities, ensuring food security and cultural preservation. These ecosystems also serve as havens for spiritual and mental well-being, offering solace amid urban landscapes. Despite their importance, rocky shores, such as coastal habitats in general, face significant threats from climate change and require legislative measures to safeguard both natural resources and cultural traditions associated with these regions.

When considering the ecosystem services and human dimensions related to rocky shores, there are many topics that should be addressed and prioritized in future research. Exploring the potential for sustainable blue economy initiatives centered around rocky shore biodiversity is paramount for fostering economic development while simultaneously promoting conservation objectives. By studying the biodiversity of rocky shore species, researchers can identify valuable resources and ecosystem services that can form the basis of sustainable economic activities. These initiatives may include ecotourism ventures, aquaculture projects focused on native species, and bioprospecting for novel biotechnological applications. Integrating economic development with conservation goals ensures that human activities are conducted in a manner that safeguards the integrity of rocky shore ecosystems and preserves their biodiversity for future generations. Moreover, by demonstrating the economic value of these ecosystems, such studies can encourage policymakers and stakeholders to prioritize their protection and management, thereby contributing to the long-term sustainability of both the blue economy and biodiversity.

Biotechnology is an area of increasing research interest in Brazil. Exploring the biotechnological potential of species inhabiting rocky shores reveals a rich reservoir of bioactive compounds and biomolecules with diverse applications in pharmaceuticals, industry, and environmental sustainability. From microorganisms to macrofauna, these organisms offer unique biochemical pathways and defense mechanisms, holding promise for the development of new drugs, enzymes, and biotechnological tools. By harnessing these resources, researchers can pioneer sustainable solutions to global challenges while fostering economic growth and environmental conservation. Collaboration between scientists, industry stakeholders, and policymakers is essential to ensure responsible utilization of these invaluable natural assets.

Engaging in citizen science projects and establishing long-term monitoring programs are crucial components of studying rocky shore ecosystems that must be supported in future research. Citizen science initiatives involve the active participation of the public in scientific research, allowing for the collection of vast amounts of data across large geographical areas that would otherwise be challenging to obtain. By involving citizens in data collection, monitoring, and analysis, these projects not only contribute valuable information on the biodiversity, ecological processes, and environmental changes occurring in rocky shore habitats but also foster public awareness and appreciation for these ecosystems. Moreover, long-term monitoring programs provide critical insights into ecosystem trends, allowing researchers to detect and understand changes over time, including the impacts of climate change, pollution, and other human activities. Citizen science projects and long-term monitoring efforts promote collaboration between scientists, local communities, and policymakers, leading to informed decision-making and effective conservation strategies to safeguard rocky shore biodiversity and ecosystem health for future generations and, thus, integrate all topics already mentioned for research.

Finally, considering the human dimension and the different knowledge systems, documenting and preserving traditional ecological knowledge associated with rocky shores is essential for both cultural heritage preservation and scientific advancement. Indigenous and local communities have developed a deep understanding of the intricate relationships between organisms and their environment over generations of interaction with rocky shore ecosystems. This traditional knowledge encompasses valuable insights into species distribution, habitat dynamics, and ecosystem functions that may not be captured through conventional scientific approaches alone. By incorporating traditional ecological knowledge into scientific research and conservation efforts, researchers can gain a more holistic understanding of rocky shore ecosystems and enhance their effectiveness in management and protection. Furthermore, recognizing the cultural significance of this knowledge fosters respect for indigenous and local communities and promotes collaboration between traditional practitioners and scientists, leading to more inclusive and comprehensive approaches to rocky shore conservation and sustainable resource management.

Most of the impacts of human activities on Brazilian rocky shores are addressed by Martinez and Altvater (2024) (Chapter 6), encompassing different drivers. Historically, fishing activities have been prominent, contributing significantly to national exploitative production, with artisanal fishing comprising a substantial portion. The exploitation of over 400 species for various purposes, including aquarium trade, human consumption, and craftsmanship, has led to population declines and endangerment for several species. Additionally, direct and indirect consequences of fishing, such as habitat destruction, ghost fishing, and the harvesting of invertebrates further exacerbate the ecological strain. Aquaculture, particularly of mollusks and macroalgae, has emerged as a response to overexploitation but faces its own set of challenges, including bioinvasions and other environmental impacts. Moreover, pollution, historically associated with industrial effluents, untreated domestic waste, and agricultural runoff, poses significant threats, introducing a myriad of contaminants into estuarine and coastal areas. These contaminants, ranging from nutrients to pharmaceuticals, have adverse effects on marine fauna and flora, altering population dynamics, reproductive patterns, and physiological functions. Given these complexities, comprehensive research is imperative to understand and mitigate the impacts of human activities on Brazilian rocky shores. This is also necessary to inform effective conservation policies and sustainable coastal management practices toward preserving biodiversity and sustaining ecosystem services.

Future research on the impacts of coastal hardening and urbanization on rocky shore ecosystems is crucial for mitigating the adverse effects of human development on coastal biodiversity. By assessing how infrastructure development and urban expansion alter the physical and ecological dynamics of rocky shores, including increasing bioinvasion, researchers can identify strategies for both mitigation and adaptation. These studies shed light on the direct and indirect impacts of coastal hardening, such as habitat loss, fragmentation, and alteration of sedimentation patterns, which can have profound consequences for the distribution and abundance of marine species. Moreover, research in this area facilitates the development of sustainable coastal management practices that balance the needs of human populations with the preservation of rocky shore habitats. By integrating ecological knowledge into coastal planning and policymaking processes, scientists can help minimize the negative impacts of coastal development on biodiversity while enhancing the resilience of rocky shore ecosystems to future environmental changes.

Future research in this area should also move on from single stressors to understanding the interactive effects of multiple stressors on rocky shore ecosystems. Additionally, there is a pressing need for long-term monitoring research to facilitate adaptive management strategies for mitigating human impacts on these environments. Future research should focus on studying species distribution in association with climate change, investigating the impacts of bioinvasions beyond mere species distribution to encompass environmental, social, and economic effects, and elucidating the quantity of catch and its sustainability challenges on artisanal invertebrate fisheries. These areas represent critical gaps in current knowledge and warrant further investigation to ensure the sustainability and resilience of rocky shore ecosystems in the face of anthropogenic pressures.

Finally, Pardal et al. (2024) (Chapter 7) contextualize scientific research in these environments, highlighting its historical importance for the development of ecology and biology. The evolution of studies, from biodiversity description to understanding ecological processes at different scales, demonstrates how Brazilian research has contributed to global knowledge about rocky coastal ecosystems. Collectively, these seven chapters provide a comprehensive understanding of the complexity and significance of rocky coastal ecosystems in Brazil, highlighting both their biological diversity and their cultural, economic, and scientific relevance. This integrated view is essential for informing conservation policies and sustainable management.

Future research on rocky shores in Brazil holds immense potential for advancing the understanding of patterns in tropical and subtropical regions in comparison with other localities. By addressing local knowledge gaps and conducting collaborative, integrated research across similar systems globally, valuable insights can be gained into broader ecological patterns and processes. Furthermore, considering the challenges posed by spatial fragmentation in current research, there is a critical need to adopt a multi-scale approach. This involves investigating processes and phenomena at various spatial scales, from local to regional and beyond, to capture the complexity of rocky shore ecosystems comprehensively. Collaborative efforts with researchers studying similar systems in comparable environmental conditions can enhance the transferability and applicability of findings, ultimately informing more effective coastal management strategies not only in Brazil but also in other tropical and subtropical regions globally.

Rocky Shore Research, Management, and Conservation

In recent decades, our understanding of Brazilian rocky shores has significantly expanded, encompassing environmental characteristics, biodiversity, ecological processes across various scales, and ecosystem services, as evidenced in the preceding chapters. However, crucial knowledge gaps still remain which impair a better comprehension of the Brazilian coastal environment and also the use of such knowledge by different stakeholders.

The enhancement of existing knowledge will not only support the continuation of scientific advance but also bolsters research groups along the coast, demanding further exploration and capacity-building, spanning intertidal and sublittoral zones, across all areas of expertise—from systematic, molecular biology and biodiversity to ecological and social aspects of rocky shores. Additionally, expanding funding support specific to research devoted to this environment, which is not competitive with other ecosystems but rather integrative, and thus ensuring adequate financial support for research groups, is paramount. By addressing these crucial areas, advancing research methodologies and fostering interdisciplinary collaborations, we can unlock the full potential of research on rocky shore ecosystems and pave the way for their sustainable management and conservation in Brazil and beyond.

To guarantee that science has a social impact, it is essential to understand the link between rocky shores and sustainable development as the relationship between science and knowledge production and decision-making. The science and knowledge about rocky shores align with the sustainability agendas, as the All-Atlantic Ocean Research and Innovation Alliance (AAORIA 2022) and particularly within the framework of the United Nations Decade of Ocean Science for Sustainable Development (2021–2030) (UNESCO-IOC 2021)—also known as the UN Ocean Decade. These intergovernmental agendas emphasize the importance of proactive and integrated action to achieve sustainable development, inviting participation from all sectors of society. The UN Ocean Decade, centered on ocean science for sustainable development, provides a unique opportunity for transformative change, recognizing the central role of science in addressing social, environmental, economic, and cultural challenges. It aims to tackle the Sustainable Development Goals (SDGs) from a coastal and oceanic perspective, emphasizing the interconnectedness of terrestrial and marine environments. By highlighting the significance of coastal areas and rocky shores, this book underscores the importance of local actions in contributing to global sustainability goals, stressing the need for individual engagement and collaboration across sectors to create a more sustainable future.

The potential for transdisciplinary research on Brazilian rocky shores emerges as a pivotal avenue for future investigation, aligning with the objectives of the UN Ocean Decade. The reflection on transdisciplinary science prompts a necessary consideration regarding the nature of scientific inquiry, advocating for collaborative efforts across diverse sectors of society. This approach underscores the importance of socially referenced scientific endeavors, which can inform effective management and conservation strategies. Moreover, the UN Ocean Decade foments an interdisciplinary understanding of ocean science, integrating environmental, biological, social, and interdisciplinary dimensions to address societal challenges. In this context, rocky shores are a key example of coastal environments that can benefit the development of transdisciplinarity. By promoting collaborative and participatory research frameworks, the UN Ocean Decade also aims to foster transformative scientific endeavors that engage diverse stakeholders and promote sustainable development. Through inclusive and dialogue-driven approaches, science can effectively address local and global sustainability challenges, positioning the ocean as a catalyst for positive change.

When considering the management and conservation, the use of the knowledge on rocky shores will be essential for a science-based decision-making. In this case, the transdisciplinary science supports that decision-makers and policymakers should already be part of the project from the definition of the question to the delivery of the results. The understanding of the governance scenarios and the communication with different stakeholders are important to guarantee that the knowledge on rocky shores goes beyond the scientific community.

It is important to understand the multilevel governance that influences coastal habitats in Brazil, as well as the management instruments that can benefit from the science-policy interface and where the use of the knowledge on rocky shores can positively impact the society. All the knowledge from rocky shores has a connection with Law proposals to current policies. There is a connection with the proposal of the Sea Law currently under consideration in the Legislative branch and a direct relationship with different regulatory frameworks, including the National Environmental Policy, National Policy on Climate Change, Environmental Crimes Law, National Coastal Management Plan, National System of Conservation Units, National Biodiversity Policy, National Environmental Education Policy, National Policy for Sustainable Development of Aquaculture and Fisheries, National Policy for Sustainable Development of Traditional Peoples and Communities, Payment for Environmental Services, National Tourism Policy, National Policy for Marine Resources, National Bioinputs Program, and Brazilian National Plan for the UN Ocean Decade.

It also relates to decision-making spaces in the government, where knowledge about rocky shores can support decision-making and the reconstruction of Brazilian environmental policies, which have been impacted over the last decade. Ocean science has historically been supported in the Ministry of Science, Technology and Innovation with a Coordination of Ocean and Antarctica. More recently, with the creation of the Ocean and Coastal Management Department at the Ministry of Environment (MMA) in 2023, marine-coastal governance takes on new processes of construction and guidelines for which multilevel governance becomes essential. For example, the MMA initiated the review of National Adaptation Plans and resumed the coastal management agenda with the recreation of the Coastal Management Group (GI-GERCO) for discussions on coastal governance, proposing the creation of a new Federal Action Plan (PAF) for the coastal zone. Moreover, it is important to consider the ongoing process of reviewing the Fisheries Law—L11959/2009—and its relationship with national fisheries governance, which assumes a new scenario with the creation of the Ministry of Fisheries and Aquaculture in 2023. Furthermore, it would be an opportunity to leverage the Legal Framework of Complementary Law 140 and, for example, implement Tripartite Committees for Marine-Coastal Governance. Thus, different governance spaces in the federal executive branch can benefit from knowledge about rocky shores in their formulation of public policies. In addition to the federal governance, Brazilian management and conservation also depend on the governance structure in the States and Municipalities. In such a case, aligning local actions, policies and governance to national actions, policies and governance, but it is a key step to guarantee rocky shore conservation based on scientific knowledge.

To promote science-based decision for management and conservation of rocky shores, it is essential to promote a transdisciplinary science with dialogue and communication of this knowledge with different sectors of the society during all scientific development. Therefore, beyond the increasing production of scientific knowledge and publication of scientific articles about rocky shores, the dissemination of this knowledge to society is essential, enhancing ocean literacy and fostering societal interaction with marine environments for engaging Brazilian society in the conservation of these environments. By delving into the cultural, educational, and societal dimensions of oceanic connections, researchers can promote a deeper understanding of the importance of rocky shore ecosystems in shaping local cultures, economies, and livelihoods. Through education and outreach initiatives, such as workshops, exhibits, and community engagement programs, individuals can develop a greater appreciation for the ecological significance of rocky shores and the interconnectedness of marine systems. Additionally, by integrating traditional ecological knowledge with scientific research, studies on rocky shores can bridge the gap between indigenous communities and scientific communities, fostering mutual respect, collaboration, and co-management of marine resources. Ultimately, promoting ocean literacy and facilitating meaningful interactions between science and society can empower individuals to become stewards of rocky shores and marine environments, contributing to conservation efforts and sustainable ocean management practices.

References

AAORIA (2022) All Atlantic Ocean Research and Innovation Alliance Declaration. https://allatlanticocean.org/wp-content/uploads/2023/03/20221307_All-Atlantic-Declaration-signed.pdf. Accessed 10 May 2024

Calado L, Violante-Carvalho N, Tâmega FTS (2024) Brazilian rocky shores: geomorphology and physical environment (chapter 1). In: Coutinho R, Christofoletti RA (eds) Brazilian rocky shores. Springer, Cham

Cordeiro CAAM, Floeter SR, Ferreira CEL (2024) Biogeographic patterns of Brazilian rocky reef fauna (chapter 2). In: Coutinho R, Christofoletti RA (eds) Brazilian rocky shores. Springer, Cham

Fernandes LDA, Da Gama BAP (2024) From benthic to pelagic environment: the benthic-pelagic pathway (chapter 3). In: Coutinho R, Christofoletti RA (eds) Brazilian rocky shores. Springer, Cham

Ghilardi-Lopes NP, Clauzet M, Barradas JI, Vinha V (2024) Economic and cultural aspects of ecosystem services in rocky shores (chapter 5). In: Coutinho R, Christofoletti RA (eds) Brazilian rocky shores. Springer, Cham

Martinez A, Altvater L (2024) Antropogenic stressors on shallow coast rocky reefs (chapter 6). In: Coutinho R, Christofoletti RA (eds) Brazilian rocky shores. Springer, Cham

Pardal A, Jenkins SR, Navarrete SA, Wangkulangkul K, Christofoletti RA (2024) Research on rocky shores in Brazil: advances and contributions to international fora (chapter 7). In: Coutinho R, Christofoletti RA (eds) Brazilian rocky shores. Springer, Cham

UNESCO-IOC (2021) The United Nations decade of ocean science for sustainable development (2021–2030) implementation plan. UNESCO, Paris. IOC Ocean Decade Series, 20. https://unesdoc.unesco.org/ark:/48223/pf0000377082

Zalmon IR, Tâmega FTS, Coutinho R (2024) Regulatory processes of biological zonation on rocky shores (chapter 4). In: Coutinho R, Christofoletti RA (eds) Brazilian rocky shores. Springer, Cham

Index

R. Coutinho, R. A. Christofoletti (eds.), *Brazilian Rocky Shores*, Brazilian Marine Biodiversity, https://doi.org/10.1007/978-3-031-67206-4

The manufacturer's authorised representative in the EU is Springer Nature Customer Service Centre GmbH, Europaplatz 3, 69115 Heidelberg, Germany. If you have any concerns regarding our products, please contact ProductSafety@springernature.com

Printed and bound by CPI Group (UK) Ltd, Croydon, CR0 4YY
28/09/2026
02224081-0018